RIVERS OF OREGON

TEXT AND PHOTOGRAPHS BY TIM PALMER

Oregon State University Press   *Corvallis*

Rivers of Oregon

The John and Shirley Byrne Fund for Books on Nature and the Environment provides generous support that helps make publication of this and other Oregon State University Press books possible.

Library of Congress Cataloging-in-Publication Data
Names: Palmer, Tim, 1948- author.
Title: Rivers of Oregon / Tim Palmer.
Description: Corvallis : Oregon State University Press, 2016.
Identifiers: LCCN 2016004824 | ISBN 9780870718502 (alk. paper)
Subjects: LCSH: Rivers—Oregon. | Stream ecology—Oregon.
Classification: LCC GB1225.O7 P353 2016 | DDC 551.48/309795—dc 3
LC record available at https://lccn.loc.gov/2016004824

∞ This paper meets the requirements of ANSI/NISO Z39.48-1992 (Permanence of Paper).

© 2016 Tim Palmer
All rights reserved.
First published in 2016 by Oregon State University Press
Printed in China

Oregon State University Press
121 The Valley Library
Corvallis OR 97331-4501
541-737-3166 • fax 541-737-3170
www.osupress.oregonstate.edu

COVER: Rogue River above Rainie Falls
i: Alsea River below Grass Creek
ii–iii: Sweet Creek above Homestead Trailhead
iv–v: : McKenzie River and Koosah Falls
vi–vii: Williamson River at Head of River Spring
vii: South Umpqua River above Days Creek (top);
    Nestucca River bubbles (bottom)
viii–ix: South Umpqua River below Winston

ALSO BY TIM PALMER

Field Guide to Oregon Rivers

Field Guide to California Rivers

Rivers of California

California Glaciers

Oregon: Preserving the Spirit and Beauty of Our Land

Luminous Mountains: The Sierra Nevada of California

Trees and Forests of America

Rivers of America

California Wild

Pacific High: Adventures in the Coast Ranges from Baja to Alaska

Lifelines: The Case for River Conservation

Endangered Rivers and the Conservation Movement

The Heart of America: Our Landscape, Our Future

America by Rivers

The Columbia

Yosemite: The Promise of Wildness

The Wild and Scenic Rivers of America

California's Threatened Environment

The Snake River: Window to the West

The Sierra Nevada: A Mountain Journey

Youghiogheny: Appalachian River

Stanislaus: The Struggle for a River

Rivers of Pennsylvania

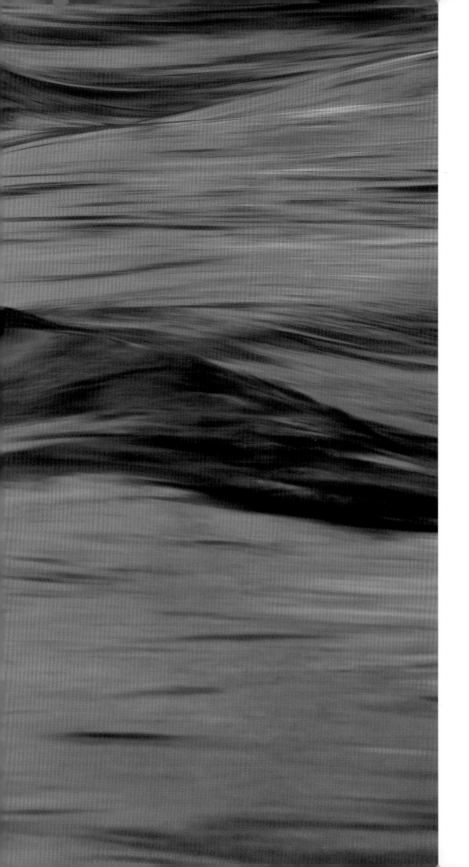

CONTENTS

1 The Essence of Oregon 1

2 Cycles of Life 9

3 The View to the River 19
   COASTAL RIVERS 21
   UMPQUA RIVER 44
   ROGUE RIVER 56
   WILLAMETTE RIVER 64
   THE COLUMBIA RIVER AND ITS GORGE 90
   DESCHUTES RIVER 100
   EASTERN OREGON RIVERS 110
   SNAKE RIVER 118

4 Rivers Through Time 137

5 The Rivers Around Us 149

Acknowledgments 157
Sources 158
About the Author and Photographer 159
About the Photographs 160
Index 161

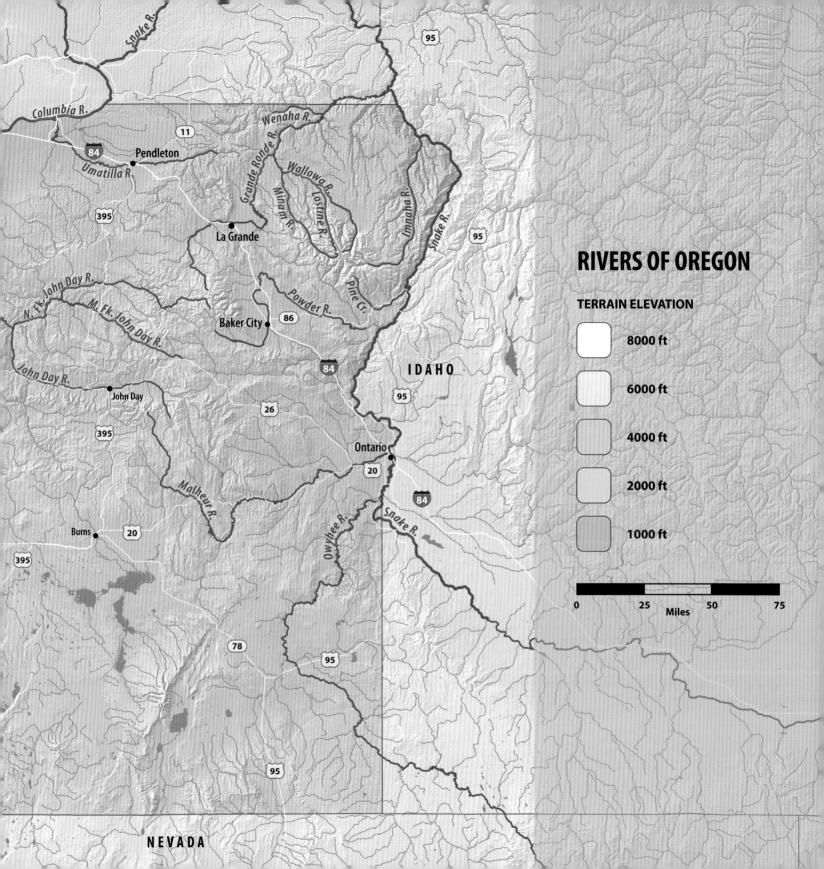

Siuslaw River

The Siuslaw rushes toward tide line upstream from Mapleton.

CHAPTER 1   The Essence of Oregon

Oregon has it all: mountains, ocean beaches, canyons, valleys, forests, deserts, lakes, wetlands, farms, towns, and cities. But foremost, Oregon is a land of rivers.

The rivers, in fact, have shaped the mountains through eons of watery wear. The ocean beaches are replenished by sand that rivers deliver to sea. Rivers carve the canyons and valleys with sinuous currents, and they nourish both forests and deserts. Lakes and wetlands are filled by rivers and then connected by them to everything that lies downstream. The farms, towns, and cities couldn't exist without water supplied by rivers.

Oregon's waterways flow from rain forests, from the Cascades' high volcanoes, from arid ranges, and from granite peaks of the Wallowa Mountains, where rapids thunder down to the Snake River at the state's eastern border. Rivers link Oregon's scenery of snowcaps with surf, pine forests with desert. They create riparian corridors that grace the state with ribbons of green.

The rivers flow from it all, and to it all, and they underpin the geography, biology, and artistry of every landscape they cross. Stunning in their beauty, essential to wildlife, vital for the workings of whole ecosystems, and critical to every person, farm, and town, our rivers are truly lifelines.

Just look! Crystal clear water reflects brilliantly all the colors of the rainbow, and the whitest of foam plunges over the blackest of jagged basalt through gorges emerald with moss, hemlock, and fir. Some windings are serene and glassy-smooth; others riffle past cottonwoods or disappear in mist at the chilled brink of waterfalls.

Words can only hint at what awaits you along Oregon's rivers. To really sense the importance of rivers, and to delight your eyes with their beauty, and to feel their hidden power, you have to see for yourself. But the next-best thing is a picture, which is why you'll find 142 photographs in this book.

As I paddled down the Willamette, with its western shore ramping up toward the Coast Range and its eastern bank climbing toward the higher Cascade Mountains, I thought of this river as the metaphorical keel of Oregon—a balancing feature that all the rest depends upon. Without the Willamette's watery path headed northward—along with the valley it shaped and the runoff it delivers—few people would ever have settled in the most populated corridor of Oregon. Portland wouldn't even exist.

I knew also that the Willamette was the supply line on which salmon, birds, beavers, and hundreds of other creatures depend. More whimsically, and more personally from my own cherished experience, the Willamette is the Huck Finn river of the Northwest. Its riffling current made it possible for me to set out on a revealing expedition of two weeks, traveling 189 miles from upper reaches down to tidewater.

My fluid route and ticketless transit touched cities, farms, and wild remnants of the original Oregon, bringing all these elements together in a sojourn that—though tame relative to other wild rivers—was spiced with adventure and discovery. I had to be alert to the swirls and boils, to currents and headwinds, to seductive back channels that drew me to explore but could serve me up to logjams or wooded swamps with no pleasant way out. But lacking steep drops—except for the largest cataract in the West at the upstream edge of Portland—the Willamette drifts at a gentle pace that's suited to many people in canoes or other craft. It invites us on a quintessential "beaver-state" outing. In my unbridled fantasies about education, paddling the length of the Willamette would be a coming-of-age ritual for every boy and girl in Oregon.

The Willamette has been affected by two centuries of efforts to tame and subdue its awesome and primordial force. But a certain magic remains, and I felt it in every view of the river bending in front of me, in the rise of great cottonwood trees at the shoreline, and even in the dark of the night. Lingering on a spacious gravel bar at my second campsite, where cold undertows of the McKenzie River joined the larger flow, I watched stars fall into a navy-blue sky. Then moonlight streaked the river in white while the current bubbled its song to my ears. From the top of a cottonwood, a great blue heron—aroused by some mystery in the dark—squawked a birdcall from the beginnings of time. A great horned owl hooted from a roost deeper in the floodplain forest, telling me that this place was his. A beaver, aware of my presence though I hadn't moved a muscle, smacked its tail on the water's surface and dove to pass by underwater.

There on the Willamette, just a few miles from Eugene, my world was reset to the millennia of days, nights, seasons, floods, droughts, and deep passages of time as the river has carved its course and as life has evolved in the stream's watery depths and along its shores. The river took me back to a universe that's lasting and real. In setting out on a paddling adventure to a place I had never been, I felt like I had come home. And I also felt that my home had grown much larger. The river opened the door to new acquaintances with everything that lay downstream.

Each day I drifted farther. I enjoyed both solace and excitement in the rhythms of river life: sunrise followed by high noon and then sunset, morning calms followed by afternoon winds, pools followed by riffles, over and over in what hydrologists call the pool-and-riffle sequence. Each evening I took my camera in hand to capture sunset's glow on animated currents. Every night a dome of stars and cottonwood limbs arced above my camp. Every morning in the tender light of dawn I watched for the wake of an otter or mink, the whistling flight of a merganser or wood duck, the dive of a kingfisher, osprey, or eagle.

The Willamette is the largest stream flowing wholly within Oregon, but it's just one of hundreds, each with its own purpose and path, its own charm and appeal, its own intrinsic worth.

I wanted to see them all.

The waters of Oregon endure a troublesome suite of problems that I describe in Chapter 4, but relative to other states, rivers here rate among the best. We still have long, free-flowing sections of streams that remain connected to the ocean without dams blocking the flow and halting the migration of fish. Waters of the Cascade Mountains and Coast Range run transparently clear in the low-flow months of summer, and you can see to the bottom where boulders, cobbles, and pebbles collect in mosaics of rocky artwork. Salmon and steelhead—extinct in many other places—still migrate up dozens of rivers to their ancient spawning grounds. At misting waterfalls you can see them leaping in the autumn. Oregon's rivers support a commercial fishery at sea along with celebrated seasons of sport-fishing on inland waters. With hundreds of miles of riverfront trails, the streamside hiking in Oregon is the best. Meanwhile, dozens of enticing whitewater runs call to kayakers, rafters, and canoeists, and we have more rivers suited for extended overnight paddling trips than any other state.

Healthy rivers are not only essential to the abundance of life and a historically robust economy in both sport and commercial fishing, but to all that we do. The livability of whole towns and regions would wither if it weren't for rivers and the water they deliver.

Oregon's rivers are likewise embedded in our history and culture, from the route of Lewis and Clark across the Northwest to urban greenways that brighten Portland, Pendleton, Eugene, Corvallis, Salem, Grants Pass, Bend, and other towns large and small. Whether in our backyards or in our most cherished wilderness, the rivers give us a refuge from the stress and clutter of our busy lives. At the stream's edge, we can adjust our expectations in synchrony with the natural world.

From the Umpqua to the Owyhee and from the Columbia to the Klamath, I've been inspired by beauty, thrilled by the rivers' motion, awed by endlessly alternating cycles of power and peacefulness. I've been fascinated by these wonders of nature. And so here, in this book, I've launched a journey to convey the enchantment these streams have offered me and to describe our flowing waters as the essence of Oregon.

Willamette River at the base of Willamette Falls
In volume of flow, this is the largest waterfall in the West, second only to Niagara nationwide. A dam has been built around the upper rim of the falls and dominates the view from above, but from the base, the dam is nearly invisible. This view of awesome power is seldom seen, even so near Portland, because—for now—it can only be reached by boat.

Rogue River near Union Creek
Snowstorms blanket the Cascade Mountains and nourish headwaters of the Rogue.

Metolius River

Near its source in one of Oregon's largest springs, the Metolius rushes north through a forest of ponderosa pines.

Chetco River, fishing for steelhead at Loeb State Park

Though anadromous fish return from the ocean to spawn in only a fraction of their past abundance, Oregon's rivers still rank among America's best for steelhead and salmon angling.

Snake River in Hells Canyon

Bill Sedivy leads the way through big whitewater of this five-thousand-foot-deep canyon that draws rafters, kayakers, and drift boaters nationwide. Dam proposals here sparked one of the great river conservation battles in American history and were halted when Congress designated the Snake as a National Wild and Scenic River in 1975.

Canoeing on Drift Creek

This tributary to the lower Alsea River winds past cutover mountainsides and then traverses the heart of Drift Creek Wilderness—one of few ancient forests remaining in the Coast Range. Canoeist Travis Hussey picks the sweet line through a rocky rapid in the twenty-mile roadless reach.

Rogue River and snowstorm

While rainfall is the main source of river flows in Oregon's Coast Range, rain and snowfall nourish hundreds of streams in the Cascade Mountains, yielding runoff that continues through dry summer months. However, the season of ample flows is shortening due to global warming; winter rain is now common where snow once consistently accumulated in the mountains.

CHAPTER 2 Cycles of Life

At the riverfront I visit often, the water always catches my eye first. Then I notice the willows, cottonwoods, cobble bars, beaches, and birds, along with the color, motion, and dazzling light saturated with green and blue by day, then gold at sunset. But beyond what I can see, the stream completes the hydrologic cycle that circulates the world's most precious fluid first from the ocean and into thin air, then down to the landscapes where we live, and finally back to sea. Because of this, rivers are fundamental to every other great cycle of life, including my own. Here I stand with over half my body weight in watery fluid—virtually all of it sourced from a river or a connected underground supply. The rivers literally flow in our arteries and veins.

With a view to those dependencies and to the related workings of the natural world, this chapter addresses the climate and geology that determine how and where our rivers run, the riparian or riverfront stage on which nature acts out its cycles of creation and transition, and the fish and other creatures that depend on the water as it flows across the land.

First, from the apparent void of the atmosphere come the essential rain and snow. In Oregon, winter storms bred by low-pressure cells migrating across the Pacific and down from the Arctic inevitably encounter our rocky coast. Eighty inches of rainfall saturates Oregon's westernmost edge in a normal year—if "normal" can be said to exist. But that's just the beginning. As the storms blow up the slopes of the coastal, Cascade, and interior mountain ranges, the temperature drops three degrees for every thousand feet of rise, and as it does, the air's ability to hold moisture weakens and the rainfall increases, topping out at two hundred inches on the windward side of some high peaks. The winter rain crystalizes to snow on Coast Range summits and whitens the higher-elevation Cascades, accumulating and traditionally consolidating to depths of twelve feet or more by winter's end. Descending air on the east side of those mountains promptly warms and regains its ability to hold vapor, and so precipitation diminishes to a mere ten inches in a rain shadow across the grassland steppe that undulates toward Idaho. Along the way, that rolling plain is interrupted by mountains that jut skyward and once again rake off rain and snow that feeds a surprising suite of desert-bound streams, all the more precious for the surrounding aridity.

Rivers wouldn't exist without this rain and snow, but the rise and fall of land is what determines where the water will run. Oregon's terrain was first shaped by earthquakes and volcanism, then refined by the sculpting erosion of water. On the state's western edge, seismic activity at the juncture of the Pacific and North American plates created the impressive rise of the Coast Range for Oregon's entire north-south length. Related phenomena to the east caused the eruption of spectacular shield and strato-volcanoes accenting the crest of the Cascades—Mount Hood in the north to McLoughlin in the south. Farther east, laminar flows of lava across the Columbia Plateau have sandwiched thick layers of basalt broken with ranges pushed up by cone-shaping volcanism or by swarms of local earthquakes. All told, the landscape of Oregon forms a magnificent template on which the waters run. Owing to these climatic and geologic differences, the rivers of each region adopt their own specialties and distinctive styles of flow. Along the coast, rivers burst from winter's rain-saturated mountains and cut deeply forested paths through soft erodible rocks that originally solidified undersea. Runoff from these coastal rivers spikes radically with winter storms, then wanes in spring and summer.

With a deeper reservoir of snowpack, and runoff that's surrendered more gradually through the melt-water months of spring and early summer, Oregon's northern Cascades send five major rivers and hundreds of tributaries down the mountains' soaking western flanks. These paths toward the Pacific are intercepted by the northbound Willamette River, which redirects runoff with a sharp right-hand turn toward the tidal Columbia. Meanwhile, the southern half of the Cascade Range slants westward from evergreen uplands of the Umpqua and Rogue basins. Oregon's fifth- and seventh-largest rivers were powerful enough to crosscut through the Coast Range while it slowly rose, and the resulting canyons now wind and stair-step toward sea. The Umpqua tours pastoral and woodland scenes with delightful rapids, while the Rogue's cleft in the coastal mountains penetrates a canyon-bound wilderness like no other.

With sparse rain on the east flank of the Cascades, and steeper slopes to negotiate, streams there have eroded shorter routes from mountaintops to volcanic plains. Northern Cascade waters foam down to the Deschutes. Like the Willamette River on the Cascades' opposite flank, the Deschutes aims directly north to the Columbia. But unlike the Willamette's gentle windings through farmland fashioned from silt delivered by rivers, the Deschutes gnaws and rips through razor-edged lava bedrock.

Farther south on the Cascades' east slope, the Klamath gathers its flow from varied headwaters and then carves a forbidding gorge that transects the entire Cascade Range. Pacific-bound, the lower Klamath careens through northern California's mountains where it negotiates seven discrete terranes or landmasses that have seismically accreted to the West Coast. At sea level it's the third-largest river on the West Coast south of Canada.

Back at Oregon's northern border, the Columbia River defines the boundary with Washington. Far upstream, this behemoth of all western waterways starts with prodigious runoff from the Rocky Mountains in British Columbia. When it lumbers across the US border, it's already larger than all but five rivers in the entire United States. Then it continues to grow, taking huge gulps from the Snake, Deschutes, and Willamette, plus hundreds of smaller streams. Nationwide, only the Mississippi, Saint Lawrence, and Ohio have greater volume. The Columbia is fifteen times the size of the more thoroughly storied Colorado River. But today the giant's true identity is hidden; throughout its length in Oregon, the Columbia is dammed. Back-to-back reservoirs down to tide line below Bonneville flood the river's entire path in the United States except for a mere fifty-two-mile free-flowing reach in Washington—the only place you can see the Columbia as an authentic free-flowing river.

Joining the Columbia east of the Deschutes, the smaller John Day River wanders in a mountain-and-canyon fastness of 293 miles from conifer clad uplands to golden rock walls downstream. In far eastern Oregon, the Owyhee River gathers springtime runoff from wild uplands in Nevada and Idaho and incises desert canyons deeper than they are wide. Northward, chilled streams swirl down from snowcaps of the Wallowa Mountains to the Grande Ronde River, which meets the Snake at the bottom end of Hells Canyon, just north of the Washington boundary.

This geography of waterways makes life in Oregon possible. Our rivers nourish sixty-nine species of native fish, another thirty-three nonnatives, and a host of other aquatic and riparian creatures. Iconic in the Northwest, four species of salmon, in dozens of evolutionarily discrete migratory runs, all depend on rivers. Along with steelhead, cutthroat trout, lamprey, and sturgeon, these fishes are anadromous, hatching in streams, drifting out to the greater food bank of the ocean for much of their lives, then returning to the protection of natal waters to spawn and die.

Reaching to virtually every major stream unblocked by dams or waterfalls, muscular Chinook salmon fin their way up rivers after the first rains of autumn send enticing runoff to the river mouths. Smaller but still a prized catch, coho salmon inhabit these and lesser streams, maturing in freshwater for a year or more before migrating to sea. Endangered sockeye salmon migrate through Oregon's reach of the Columbia to lakes of Washington's Wenatchee basin and to the Salmon River of Idaho. A fourth salmon species called chum just barely survives in the streams of Oregon's north coast.

Salmon are keystone species, meaning that a whole ecosystem of other creatures depends on them. Forty-one mammals, eighty-nine birds, five reptiles, two amphibians, and a host of insects and other invertebrates feed directly on salmon. Even more important, the fish transport nutrients—in the convenient form of their body flesh—from the ocean, where such elements are plentiful, far upstream to landscapes where nutrients are scarce. This life-enhancing by-product of the salmon's reproductive urge was for eons vital not only

Chinook salmon in the Rogue River at Rainie Falls

Salmon, steelhead, and other anadromous fish hatch in Oregon streams, migrate to the ocean for much of their lives, then return to their natal streams to spawn. This wide-ranging life cycle is dependent on clean, dam-free rivers from headwaters to sea.

along the water's edge where the fish rot after spawning, but also across wide swaths of bottomlands where bears and other predators caught, carried, and ate salmon, then left the remains of their prey. Historically the fish migrated upstream in such abundance that they literally fertilized an entire landscape, giving rise to towering riverfront forests, fruitful riparian gardens, and the energized health of complex communities from microbes to moose, bears, and whole Indian tribes. In recent decades, however, the numbers of fish have dipped dangerously low, virtually eliminating the age-old nutrient supply, with far-reaching consequences that we are only beginning to realize.

Especially prized by sport anglers, steelhead are rainbow trout that migrate to the sea and bulk up in the ocean's greater nutrient pool. Unlike salmon, they survive their first spawning run and return to sea for another shift of growth.

Smaller than steelhead but splendidly beautiful with black-spotted, yellow bodies, cutthroat trout can either stay

Roosevelt elk along the Umpqua River
This largest species of American elk can be seen at the Bureau of Land Management's Dean Creek area west of Reedsport. Beavers, minks, and otters are more often recognized as river-dependent mammals, but even these big ungulates need the healthy flows that maintain their floodplain and wetland refuge.

in rivers or migrate to sea, but even then they don't wander far offshore. Heavily dependent on river quality and flow, cutthroat symbolize the biological and hydrological connections between forests, rivers, and ocean.

Also anadromous, sturgeon historically grew to colossal lengths of sixteen feet—America's largest freshwater fish. These giants still swim in from the ocean to lower reaches of a few large rivers, but their migrations upstream and down have nearly been eliminated by dams.

Unlike anadromous species, "resident" fish remain in freshwater all their lives. Here we find rainbow trout that do not venture to sea as do steelhead, some cutthroat trout, whitefish, suckers, and also rare endemics that are found only in limited areas. For example, Umpqua dace and Umpqua chubs appear only in the Umpqua River. Feisty bull trout thrive in wild basins mostly east of the Cascade Mountains. Lahontan trout endure as relics of the Ice Age; eighteen thousand years ago they inhabited an inland sea of glacial runoff that flooded a large part of today's south-central Oregon desert. Now, only a few survive in arid, landlocked basins. Resident fish also include a host of nonnatives, such as bass, pike, catfish, sunfish, and carp, most of which diminish and threaten native species.

Fish are difficult to see, but a host of birds are plainly evident as they feed along the water. Mergansers—the most common ducks on rivers—dive for food and wing their way upstream or down like little choreographed squadrons of jets. Spotted sandpipers peck for insects on sandy flats. Ospreys drop from great heights to snatch fish in their talons, while kingfishers rattle their call and plunge for quarry from perches on trees or rocks. Bald eagles roost on top of tall trees and watch for a meal to appear below. Dippers sing like virtuosos in shady recesses and probe for insect life in quick currents and waterfalls.

Beavers, minks, muskrats, and otters are all river-dependent mammals, but nearly all vertebrates benefit from water, riparian woodlands, and wild corridors that buffer streams from disturbances nearby. According to the *Oregon Wildlife Diversity Plan*, riparian or riverfront habitat is essential to 55 percent of the state's amphibians, 66 percent of the reptiles, 49 percent of the mammals, and 40 percent of the birds. Far more than these—in fact, most wildlife—use rivers at one time or another.

The life of rivers depends not only on water that's clean and plentiful, but also on patterns of natural flow established through the ages, including floods. Though we often think of them as disasters, or at best nuisances, high flows in fact are essential to the life of the stream. Big water that rises with coastal storms or the rush of springtime snowmelt fills the riverbed and overflows with accumulating power in rolling waves—a foaming force that's frightening if you're in the wrong place, and at the wrong time, but spectacular and thrilling if you're not.

In their roiling flows, floods sculpt and renew the shape of the riverbed in a process that's essential to what ecologists call a "disturbance ecosystem." The river's proper functioning depends on this periodic disruption even though the cycle might appear a bit painful. For example, both pools and riffles are critical to river health—pools for cold water and shelter sought by fish, riffles for enriched oxygen and for shallow rocks as habitat for invertebrates and other life. But pools and riffles don't just happen by themselves. They're made by floods. Though we associate pools with slow water and riffles with fast water, these features ironically result from opposite phenomena. It's high flows that scour out the pools. Imagine a focused jet of water—like a fire hose aimed directly at the dirt in front of you—strong enough to push sand, gravel, and rocks away and dig a hole in the ground. In the river channel, similar action during floods blasts out the cavities that become pools during lower flows. Likewise, riffles and rapids are

Elk River in flood stage

Floods are often regarded as troublesome because we've built houses in harm's way, but big waters are essential to river health. The high flows maintain important and varied riverbed habitats by scouring out pools and also forming riffles. Floods nourish forests that shade the streams and shelter a whole community of life, and they recharge groundwater that later seeps back into the stream during months of low flow. High water makes floodplains accessible to fish for shelter and foraging, and it washes organic detritus into the stream to become food. All these values, plus recreational access, are lost when floodplains are developed, paved, or farmed.

formed where the flood flow pauses enough to deposit the densest, heaviest, and hardest-to-carry boulders. Next it deposits lighter rocks, then smaller cobbles, and together they form rapids.

While the flood's rearrangement of the riverbed might appear destructive in the short term, it underpins a structure that's essential to river health in the long term—the "pool and riffle sequence." Without floods, riffles erode and pools fill with silt, all trending toward a constant glide of undifferentiated current lacking variety. But variety is essential to aquatic life. Its loss occurs when upstream dams eliminate downstream flooding, when straightening and riverbank riprap make the channel uniform, and when chronic low flows caused by diversions allow vegetation to encroach on the streambed.

The creative aspects of floods are also crucial to riparian forests. Consider cottonwood trees. These largest of riverfront hardwoods are elegant, girthy, muscularly limbed, and shapely as they reach for the sky. They cast precious shade, stabilize riverbanks with tangled roots, and sponsor a sheltering habitat for a whole cast of birds and mammals that live in tree-trunk cavities or seek branches and exposed snags for perching and nesting. The leaves, buds, bark, and twigs are a smorgasbord for wildlife.

These trees don't just reproduce in an ecologic or hydrologic vacuum. Rather, their germination depends on periodic flooding. Genetic programming has billions of wind-blown seeds ripening soon after waters recede in late spring. Floods are essential because the seeds require either a freshly scoured surface or a new deposit of sand

South Umpqua River cottonwood forest

Cottonwoods are the largest hardwood trees along Oregon's rivers. A keystone species, they provide shelter, nest sites, and food for many birds and animals. To reproduce, cottonwoods need periodic flood flows that deposit silt on the floodplains and enable new seeds to germinate.

North Santiam River and white alder forest

Floodplains are frequently swept clean of fallen leaves and other nutrients providing nitrogen, which is essential to plant growth. But thriving in the roots of alders, microbes are able extract nitrogen from the atmosphere—where it's plentiful—and convert it into a form consumable by the tree, ultimately benefitting the entire riparian forest ecosystem.

or silt to take root. Without floods, the scouring and depositing fail to occur, and so the seeds fail to find hospitable homes. Cottonwood trees seldom live more than a century; in rivers where dams have checked floods, the forests needed by wildlife and fish are dying out, and they won't be replaced.

While our attention to floods is often fixated on the human drama of high-water hazards, those problems would scarcely exist if we hadn't built homes, buildings, and roads in harm's way—squarely on floodplains where we know—or at least should know—that the water will inevitably rise. Rivers have always flooded, and will flood even more with the intensified weather caused by a changing climate.

The great wildlife biologist and ethicist Aldo Leopold said that when nature looks good to our eyes, it's usually good for the life around it. "A thing is right when it tends to preserve the integrity, stability, and beauty of the biotic community. It is wrong when it tends otherwise." Acknowledging the importance of beauty is an unusual and courageous thing for a scientist such as Leopold to do, though it's an everyday admission for a photographer like me. Seeking out beauty endlessly but also striving to understand the workings of nature, I couldn't agree more with Leopold. Exploring and hunting for beauty along Oregon's rivers, I've concluded over and over that when it *looks* good, it *is* good.

The translucent green pools, for example, catch my eyes with their color, and those pools are also essential for trout and salmon that depend on deep water that stays cold even through hot summer months. The spectacular rush of whitewater is a sight that pulls on my camera like a magnet, and it's also vital for boosting the water's oxygen content—essential to all aquatic life. While colorful bars of rocks, cobbles, and gravel draw me to focus on those gem-like collections, the undersides of those rocks make homes for caddisflies, stoneflies, and mayflies—fundamental to the chain of life. Ancient conifers at the water's edge hold me rapt with a sense of wonder and send me scrambling for adequate perspectives through my wide-angle lens, and they are also needed for the shade they cast over the water. The sequence of riffles, rapids, and pools define beauty for me, and they also offer variety and habitat needed by the entire streamside community.

Touring the rivers of Oregon, and searching with my camera for remarkable scenes along these waters, I am drawn again and again to the features that make rivers healthy and to all the great cycles of life.

Elk River and Grassy Knob Wilderness below Bald Mountain Creek

To protect the Elk River's salmon and steelhead, forester Jim Rogers and Jerry Becker of the Elk River Land Trust persuaded Congress to designate the Grassy Knob Wilderness in 1984 and Copper Salmon Wilderness in 2009, safeguarding the river's north side and headwaters and making the Elk one of the best-protected watersheds on the West Coast.

CHAPTER 3 The View to the River

Something about the sight of flowing water simply thrills me. Of course rivers are beautiful with their windings and whitewater plunges, with their silky foam and bursting bubbles, with their colors reflecting everything around them and also revealing dark shades of the underwater world with its vivid animation and hidden mysteries. But there's more than all of that.

The river moves, and I'm drawn to its arrival and its passing. Watching the flow, I'm tugged toward the river's source, curious in the same way that I'm lured to other creation stories—actual or mythic—revealing origins of other kinds. No less than this, I'm pulled toward the water's destination. Let's go! The simple act of staring downstream is, for me, prerequisite for stepping into my canoe or raft, kicking off from shore, and welcoming the great unknown.

Beyond the water, I see the banks, beaches, flooded shorelines, cobble bars, rock piles, riparian thickets, and waterfront forests, along with views of mountains, hills, canyon walls, and a big blue dome of sky, all with a perfect foreground of river. Looking for new perspectives, I've floated, waded, and sat in the water to take pictures. I've scrambled up hillsides, climbed trees for top-down panoramas, and dived as deep as I could go to see bubbles and bottom-life from a fish's-eye view.

It's important to know a river, along with the creatures it shelters, the life-support systems it nourishes, and the practical gifts it grants to us all. For me, simply seeing the river is the first step toward knowing it in all those ways. That's one reason I love to take pictures—to capture exactly what I see in this stepping-off place toward a better understanding of the river, and of the earth, and of my place on it.

Exquisite scenes of Oregon's rivers can be discovered along hundreds of waterways where roads, paths, or brushy scrambles lead to the water's edge. Arriving there can be a rewarding experience with an enticing view and a chance to capture the beauty and nature of the stream with a camera or simply with a memory. That's all good. And it's enough. But traveling on a river opens up a whole new world, with an enhanced perspective that comes only by joining the flow and traveling with the current. I see new shoreline delights around the next bend, through the next mile, and continuing on the course of a week-long journey. River travel is guaranteed to reveal a world of wonder to anybody who is open to the sights, the scents, and the sounds of the waterway.

On some rivers, going with the flow requires well-honed paddling or rowing skills, but on many other rivers it can simply mean stepping into a raft, canoe, or kayak with some nominal training and experience, plus an instinct for caution and willingness to learn. Short trips of a few miles can reveal a lot, but several days on the water show more. After a while, it feels less like visiting and more like truly belonging—like *living* there. Some trips take on the nature of an epic journey from sources where you might barely be able to float your craft, down to broad rivers pushing toward sea or to big waters of the Columbia, Snake, or Willamette. Oregonians are fortunate to have more rivers suitable for long river trips than almost any other state—ten rivers can be run for nearly one hundred miles or longer without being stopped by dams. In write-ups accompanying the photo galleries that follow, I highlight some of those longer river trips.

The photographs that follow are organized from Oregon's Pacific slope and eastward to the Snake River, first moving through the Coast Range from the Columbia to California, then through the Cascades, then through the mountains and canyons of eastern Oregon. While these are also rivers of culture and history, my images gravitate to nature—to the original Oregon, or at least to some semblance of it. The natural world is with us still, and we can see it along these rivers.

Flora Dell Creek

Illumined by one beam of sunlight entering its cool shady alcove, basalt bedrock glistens at the foot of a tributary waterfall in the lower canyon of the Rogue.

Coastal Rivers

The Siuslaw is one of the longest rivers nestled solely in the Coast Range, so I was eager to learn what it would reveal about the watery paths that streams take through the rain-soaked mountains facing the Pacific. I wanted to float as much of the river as possible—to launch as far upstream as I could while spring runoff remained strong. So I planned my trip for April.

Hearing that the rapids were rated a relatively easy Class 2 from Siuslaw Falls down, my paddling buddy Travis Hussey and I decided to launch a four-day, seventy-seven-mile expedition from there. The weather forecast was good, and we had little reason to expect anything but a carefree cruise to tidewater.

A morning chill lingered in shaded canyons as we drove upstream from our homes at the coast, but the air carried the fertile scent of springtime's damp soil while tributaries bubbled at every turn. The first buds of the year were bursting into yellow-green leaves. With rivers running

Siuslaw River at rapids above Swisshome

high—but lacking the edgy push of winter storms—spring is a great season to explore Oregon's rivers, especially in the temperate Coast Range.

We drove upstream on the remote and narrow paved byway, close to the river for much of its length. But trees blocked views to the water, so we didn't see it much until we arrived at Siuslaw Falls. From a bridge just downstream, we finally got a good view and saw that a log immediately spanned the narrow flow and blocked our route. Hoping for the best, we drove back downstream and put in at the next opportunity. Floating free, we paddled away with the thrill of finally being on the water. But around the next bend we encountered a larger logjam. Then another. And another.

Siuslaw River logjam
Logjams can be expected in most small rivers of Oregon's forested regions. These complicate downriver boating but are valuable for fish habitat and the health of the streams.

Before the second day was over, we had dragged our canoes over eighteen logjams. Some were just one trunk wide but still required us to gingerly step out and tug on our loaded boats. One logjam was as big as a house, and we threaded a path both over and under massive boles in the prickly pile that had taken on the bearing of a whole landscape. The river purled darkly underneath, giving us pause with each step, and we took care to not miss the mark, or slip, or depend on a log that would sink or break. Crumbling banks choked with barbed salmonberry bushes offered few options for portage.

The logjams challenged us, but they also disclosed the true nature of a small Oregon river in deeply forested country. Before the logging era, plentiful logjams were the norm—surprisingly important for habitat and maintenance of a healthy river, which we'll see in Chapter 4. And as our trip progressed, negotiating the logjams turned a fairly uneventful upper Siuslaw outing into a great adventure that we both relished, right there in our own Coast Range backyard. "How are we going to get through *this* one?" I yelled to Trav more than once.

In other places, boulders had been placed across the river by the Bureau of Land Management in an effort to restore some of the channel complexity that was historically created when massive logjams ponded the flow, replenishing groundwater, lowering water temperatures, and accumulating spawning gravels. Boating on a fairly high flow, Trav and I found slots in these "cascades" where we could slip through, but we lined one congested drop by scrambling to shore and remotely guiding our loaded canoes—with ropes attached to bow and stern—through the unrunnable maze.

The logjam difficulties faded behind us after we passed BLM's Clay Creek Campground late on day two. From there the Siuslaw became an amiable forested cruise for twenty-six miles to Richardson access below the Highway 126 bridge. Tributaries bolstered the flow. We slalomed

Nehalem River below Nehalem Falls

As Oregon's longest river exclusively draining the Coast Range, the Nehalem begins on an agricultural plateau in the northwest corner of the state and loops north, west, south, and then conclusively west to the Pacific through lushly wooded mountains. Renowned for salmon and steelhead diversity, the Nehalem also offers excellent rapids for whitewater paddling and a steep cascade at Nehalem Falls. In winter, runoff can crest at thirty thousand cubic feet per second—comparable to the Grand Canyon of the Colorado but from a watershed that barely extends forty miles inland from the sea.

Nehalem River and eddy
Below Nehalem Falls' main pitch, the current spins counter-clockwise in an eddy formed by a band of bedrock jutting into the stream.

through bubbling rapids, delighted with the joy of whitewater that offered little risk to our boats weighted with camping gear.

In the 1980s we would have seen clear-cut logging all along this route, and we still did on the checkerboard of industry-owned land where timber harvest remains poorly regulated by the state. We noticed shorelines where the nominal required buffers clearly appeared to be violated, and others where the setbacks proved inadequate to avoid landslides and blowdowns of the thin strip of trees that had been left.

However, stands of enchanting old-growth forest remained in some of the pockets of public land we passed. After decades of overcutting at rates that could never be sustained, and following a contentious era of protest, controversy, lawsuits, reform, and finally insightful leadership by supervisor Jim Furnish, the Siuslaw National Forest in the 1990s reinvented itself with the goal of restoration. Clear-cutting was declared off-limits, not only along the riverfront but throughout the forest. Thousands of miles of unneeded logging roads—otherwise bleeding silt into tributaries and triggering landslides that ruin salmon and steelhead spawning beds—were closed. Now, on the recovering Siuslaw, you can drift through forested canyons and imagine a future that's better than the past. Hopefully the forward-thinking measures for protection will survive inevitable political challenges.

With daylight fading on our third day, Trav and I found a camp spot big enough for two tents, and as the evening grew dark and ominously cold, we quickly fired up dinner and then battened down for the night.

The next morning an inch of snow covered the ground. Spring had reverted to winter, and the three most difficult rapids of the trip lay ahead on our final blustery day. With flurries blowing, I carried around the largest drop rather than risk a frigid swim with my loaded canoe. Trav lugged his gear on my portage route and then flawlessly paddled the obstacle course in his lightened boat.

The Siuslaw doubled in size at the Lake Creek confluence and instantly became a big river. It pushed us over wide ledges of whitewater, through jets of current, and into whole inland seas of choppy flow with an arcing sweep of puffy-clouded sky overhead and a long fetch to shorelines on either side. Our little Siuslaw—first a headwater gauntlet of logs and then a sweet melody of winding woodland riffles—now became a pulsing Northwestern artery, full and powerful as it built toward sea.

In eight more miles the current slackened at tide line with terminal eddies spinning against brushy banks. A mile later we landed at the Mapleton dock, satisfied with a trip that had crossed the width of the Coast Range and given us a view of Oregon that could only be seen from a river.

Salmonberry River

Little-known, roadless, wondrous, the Salmonberry is a premier wild river of the Coast Range. It can be seen by walking along an abandoned railroad upstream of the Nehalem confluence.

Wilson River and tributary

The lower Wilson draws anglers to a popular fishery while mid-reaches offer the Coast Range's best whitewater within a short drive of Portland. Though Highway 6 follows the same route, steep canyonsides buffer traffic and leave the corridor of this steel-blue river feeling wild. On a cool winter day between rain showers, I caught this photo two miles below Kansas Creek.

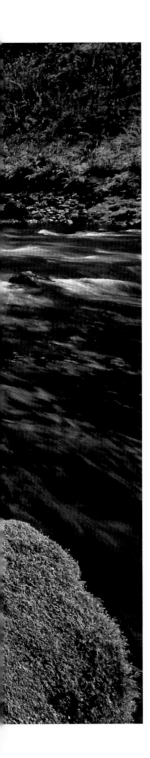

Kilchis River and old-growth forest

The Kilchis rises among emerald canyons with lucid water and mossy shores east of Garibaldi. One of the region's most productive salmon streams, it's managed as a wild fishery and also offers excellent whitewater paddling in winter here above Kilchis County Park.

Salmon River at H. B. Van Duzer State Park

This view from the water shows old-growth Sitka spruce overhanging shaded banks. In the 1930s the state protected twelve miles of the Salmon in one of Oregon's earliest conservation efforts focused on a river corridor.

Nestucca River at Alder Bend Campground

The Nestucca drops from intimate headwaters reached by a narrow paved road to this wooded campground. Farther downstream, the gently riffling river is popular with salmon and steelhead anglers rowing drift boats.

Siletz River

The striking whitewater of the upper Siletz above Moonshine County Park churns through industrial forest tracts, and below Logsden, gentler riffles and reflective pools make this river a pleasant float. The lower Siletz remains one of the most diverse salmon and steelhead streams on the north coast.

Alsea River below Salmonberry Road

Reached from ramps and paths along Highway 34, the Alsea flows through riffles and a few small rapids to tidewater. Used by fishermen in drift boats in winter, the Alsea also makes a fine canoe trip in spring while big-leaf and vine maples brighten the shores.

Siuslaw River estuary

The Siuslaw originates only two miles west of the Coast Fork Willamette and runs 109 miles to sea—Oregon's third-longest dam-free river flowing into the Pacific (only the South Umpqua/Umpqua and Nehalem are dam-free and longer). This aerial view shows the Siuslaw estuary east of Florence. Though channelized in the foreground, wetlands and sloughs on the south side remain intact, nourishing young salmon that rear in nutrient rich water.

Cummins Creek

Cummins and nearby Rock Creek are the only sizable streams on the coast with no valley roads or development, except for Highway 101 bridges. Most of their watersheds are protected as wilderness. With the neighboring Big, Cape, and North Cape Creeks, this cluster of streams nurtures native coho salmon, steelhead, and cutthroat. The gentle rush of water over spawning gravel in this photo is reached from a path off the Cummins Creek Trail.

Sweet Creek

Oregon is a state of waterfalls, and Sweet Creek drops through a stunning chain of them on its short, steep path to the ocean, seen here above Homestead Trailhead southwest of Mapleton.

Tenmile Creek, south of Reedsport

Along Oregon's 310-mile coast, twenty-six rivers enter the sea. Though the mouths of most of these and of many smaller streams are altered by channelization, riprap, dredging, and jetties, a few remain natural, including this more southerly of two creeks named "Tenmile." The bird's-eye view shows upstream dune-barrier lakes that provide coho habitat. Below, the creek winds through sand dunes and finally confronts ocean breakers. Dunes such as these blocking streams' direct exits to sea—and forming low-elevation freshwater lakes in the process—are common along Oregon's central coast.

Smith River above Smith River Falls

Splendidly isolated, the Smith River links the crest of the Coast Range to the Umpqua estuary north of Reedsport. The watershed's mix of industrial forest and public land has been heavily logged, yet a wooded riparian corridor remains for much of the ninety-one-mile length, accessible by a quiet paved road that offers one of Oregon's finest riverfront bicycling routes. Rarely paddled, the river has semi-wild shores with few logjams, and, in the spring, outstanding intermediate rapids. Restoration of this little-known gem has become a goal of Trout Unlimited.

South Fork Coquille

In the acreage of its watershed, the Coquille is the largest river basin located entirely within Oregon's Coast Range. Here, upstream from Powers, the South Fork thunders through a magnificent rain-forest canyon with steep rapids and riverfront stands of tall trees.

Sixes River at mouth

At one of the most natural and spectacular river mouths along the Oregon Coast—or anywhere—the Sixes River appeals to harbor seals and western gulls due to the biologically rich mixture of fresh and salt water found wherever rivers enter the sea.

Elk River below Butler Bar

The Elk River is known for its clarity and prized Chinook and steelhead. Nine miles east of Highway 101 a paved road enters the Elk's old-growth canyon and continues upstream along the river's steep, forested path.

Panther Creek and salmon spawning habitat

On this Elk River tributary, low gradient "flats" formed by past logjams have left gravel deposits that now serve as prime spawning grounds for salmon and steelhead. The north side of the Elk has been protected as wilderness, but south-side basins including Panther and Bald Mountain Creeks remain vulnerable to clear-cut logging on steep slopes. To safeguard fish habitat, local conservation groups have proposed protection of those tributary watersheds.

Hubbard Creek at the Pacific Ocean

With seventeen-hundred-foot Humbug Mountain in the background—tallest peak rising directly from the Oregon coastline—Hubbard Creek riffles in the warm glow of sunset through sand bars to its mouth at the sea.

Chetco Gorge

Seismically warped formations of bedrock create the Chetco Gorge, seen here at Conehead Rapid in the deepening blue of twilight.

Chetco River below the South Fork

As the wildest major river on the West Coast south of the Olympic Peninsula, the Chetco has no dams, no roads along its upper thirty miles, and little development until near sea level. Half the length lies in the Kalmiopsis Wilderness and three-quarters is designated a Wild and Scenic River through national forest. The upper stream's boulder-riddled rapids widen to a lower basin where dreamy blue pools delight swimmers and campers in summer. In autumn and winter, runoff swells and salmon return to spawn.

Umpqua River

Flowing across western Oregon from the heights of the Cascade Mountains and then through the Coast Range to the ocean, the Umpqua is a river of superlatives hidden among common views of water, forest, and countryside. This is the state's most diverse fishery, supporting multiple year-round migrations of salmon and steelhead, a lot of introduced bass, and several endemic fishes found nowhere else. The main stem is the largest dam-free river emptying into the Pacific south of Canada, and the combined South Fork and main stem are the longest—free flowing for 226 miles. Many access areas welcome boaters, anglers, and swimmers, while riverfront roads make great bike routes.

Not just a great Oregon canoe trip, but a great *American* canoe trip, the entire main stem Umpqua can be paddled throughout the low-flow months of summer and autumn—boatable seasons on only a dozen of Oregon's larger rivers. The Umpqua is not a wilderness outing, but a tour of pastoral Oregon, with ranchland above the shores, cabins along country roads, and industrial forests clearcut on raggedly steep slopes that—once destabilized by over-cutting—await the next landslide. Yet the river and its world remain remarkably beautiful. Occasional tracts of older timber stand tall and shaggy on ridgelines, indicating where federal Bureau of Land Management parcels have so far escaped the chainsaw. Scattered and rare, those enclaves of public land are threatened by proposals to loosen logging restrictions on the relatively few remaining uncut parcels in the Coast Range.

In the welcome heat and grassy scent of July, I launched a week-long expedition from the beginning of the main stem—just below the North and South Umpqua confluence near Roseburg—to saltwater. Long quiet pools led to hundreds of small rapids and a few larger drops that required me to eddy out in my canoe, scout, and strategically plan my moves or consider portages. Near the end, above Scottsburg, I carried my canoe around the flume of whitewater called Sawyer's Rapid.

The fishing seasons for salmon and steelhead were long past, and the Umpqua does not attract the whitewater aficionados of the Rogue or Deschutes, so I saw few people on my summertime sojourn. Here, in a place remote yet civilized, I could paddle for epic mileage through western mountains and feel connected to a force that I simply call "riverness." I floated contentedly for hours and days. It was just me, and the flow, and the life of the stream—a merganser diving here, an eagle soaring there, a sleek furry mink working the shore, a stout beaver slapping its tail and diving for cover.

After a week of paddling and camping, I entered the tidal, fjord-like cleft between high emerald mountains where the first scent of the ocean hung in the air. My trip ended a day later at Reedsport, where Oregon's fifth-largest river broadens in a windswept estuary before disappearing into the sea. The Umpqua had shown me a cross section of Oregon, from the Cascade Mountains to the Pacific.

Umpqua River below the North and South Umpqua confluence

North Umpqua River at Apple Creek

Renowned for summer steelhead, the North Umpqua is one of the most beloved Cascade Mountain Rivers. Its choice fly-fishing reach above and below Steamboat Creek has drawn anglers nationwide for a century. Other sections offer great angling and superb whitewater boating. With summer-long flows and a hiking and mountain biking trail seventy-nine miles long, this is Oregon's premier road-accessible recreational river.

North Umpqua and Tokatee Falls
The upper North Umpqua drops over this eighty-foot outcrop of hexagonal basalt reached by a short trail.

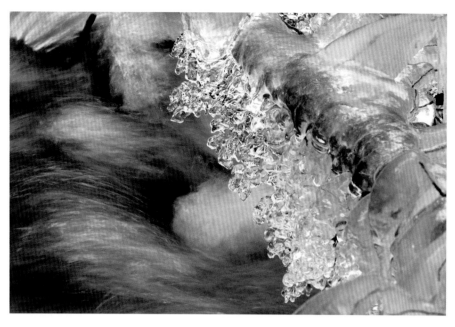

Clearwater River
In winter's deep freeze, ice lacquers fallen branches over this North Umpqua tributary.

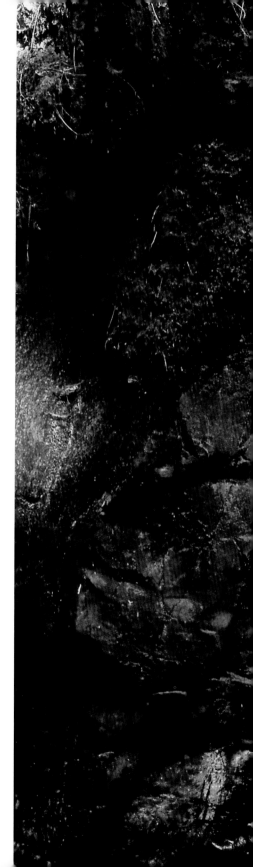

Susan Creek Falls
One of many wonders along the North Umpqua's 106-mile descent from Cascade Mountain snowdrifts to the main stem near Roseburg, this tributary hides in recessed volcanic bedrock along a path reached from Highway 138.

South Umpqua below Boulder Creek

Draining headwaters that are lower in elevation than the famed North Umpqua, this sister stream carries less runoff and quickly fades to rocky levels in summer. Downstream it skirts a lineup of towns in the I-5 corridor, yet the river remains a lifeline for fish and for recreation in southern Oregon. Lower reaches are considered the state's most productive smallmouth bass fishery, and winter steelhead draw drift boaters from January to March.

South Umpqua above Days Creek

From Teller down, the South Umpqua flows mostly in delightful riffles and small rapids, making it possible to combine a seventy-six-mile South Umpqua float with the main stem Umpqua for a total of 176 miles to Reedsport with only a few short portages—one of the outstanding extended canoe trips of the West. This turbulent whitewater chute—easily portaged—lies two miles above Days Creek.

Cow Creek headwaters

From a trailhead east of Galesville Dam, a wooded path drops down the East Fork of Cow Creek for half a mile and then winds up the main stem through exquisite low-elevation old-growth forest, fording the stream often.

Cow Creek at Island Creek recreation site

Cow Creek cascades at BLM's Island Creek recreation site, southwest of Riddle. This stream is one of the largest draining the east side of Oregon's Coast Range.

Cow Creek at Buck Creek Bridge

Below its headwaters reach, the little-known Cow Creek—more descriptively called Coast Fork of the Umpqua—loops for thirty miles with no development except a narrow paved road alongside. Industrial forest and Bureau of Land Management tracts here have been systematically clear-cut, yet a sense of wildness lingers—and could be restored—along this small river.

Rogue River

Steeped in western lore, the Rogue from Grave Creek to Foster Bar churns through thirty-four miles of captivating wild canyon. Dozens of lively rapids, seductive campsites, and hotspots for hooking salmon and steelhead make this Wild and Scenic River one of the classic float trips in the West.

To see even more of the distinguished waterway, Ann and I began our expedition at the base of Savage Rapids Dam, thirty-three miles upstream of the usual put-in. That antiquated labyrinth of cracked concrete and rusty bolts was removed in 2009; the dam was no longer needed for irrigation, and without it, salmon that had been hindered for decades were free to reach fruitful spawning grounds above. Two other main-stem dams farther upstream have been removed since we embarked on our extended Rogue journey, but at the time Savage Rapids was the highest we could go without being blocked by dams.

A few miles downstream we waved to kids along the shores in popular parks at Grants Pass. Then we continued through languid pools and babbling riffles to the first big whitewater at Hellgate Canyon. Another day and fourteen miles of frisky waves led to Grave Creek and the bedrock gates of the Rogue's wild canyon.

Miles of rapids and pools followed, beginning with the steep drop at Grave Creek "Falls." Poised at the brink of the four-foot pitch in a frozen instant of time, utterly powerless to change my fate, I felt like I would be pushed by the vertical force of the drop into the bottom of the river. But I buoyed up in big waves with hardly a drop of water in my raft.

At Rainie Falls we avoided the hazardous main channel and instead pin-balled down a rocky chute on the far right. Blasted out decades earlier as a fish ladder to coax salmon upstream, it now doubles as our sneak-route downriver.

Two days of contentment later we rode boiling water through the improbably narrow Mule Creek Canyon—a cleft twenty feet wide in places. At the Coffee Pot the current stopped but the turbulence didn't, and our downward momentum instantly transferred to a static boil underneath. The raft seemed weightless, supported on air as much as water while bubbles percolated up, hissing across the river's surface. Bobbing between rock walls, I felt like we had reached the crux of the Coast Range—the spot where our westbound route by river precisely intersected the north-south axis of mountains that continue with infrequent low breaks for thirty-six hundred miles between Baja and the Kenai Peninsula of Alaska.

Then came Blossom Bar, a consequential rapid demanding heightened focus. If you fail to pull right at the crucial instant, you're swept against an unforgiving jaw of canines, molars, buckteeth, and fangs that the Coast Range snarls in your path. I made the cut, and Ann gracefully followed with just a stroke or two in her kayak, and after another hundred yards of breaking waves and rock dodging, we each let out a jubilant shout.

In another eleven miles, at Foster Bar, most people disembark from a trip that calls them back again and again to the Rogue's varied recipe of relaxation and excitement, comfort and hardship, hot days and cold swims, all within easy reach of anyone between San Francisco and Portland. But wanting to see it all, Ann and I waved to the retiring crews deflating rafts on the ramp, and we kept going.

After a few more rapids the lower Rogue opened into a wide canyon with wooded slopes and a road perched high enough that we rarely heard traffic. Cobble bars grew vastly larger to an Alaskan scale of water and rock,

Rogue River in its canyon below Grave Creek

as if a glacier had bulldozed through less than an ice age ago. But it was flooding, such as the infamous Christmas Day torrent in 1964, that shaped the riverbed in a brand of beauty created only by the broad paintbrush of natural disturbance. A few powerful wave trains eased into glassy riffles and a peaceful drift.

But not as peaceful as it might have been. Throughout the lower Rogue below Blossom Bar, jet boats carried rows of auditorium-seated tourists and roared upstream and down, loudspeakers functioning well. The drivers courteously waited for us when they needed to, and we all waved back and forth, but time after time we stroked toward shore to let them by, bobbing on wakes that pounded the banks. In midafternoon, winds blew us to shore for a shift of enforced leisure; I read a book and Ann painted watercolors. Through it all, the lower Rogue had assumed the majesty and force of a mature river, and we began to anticipate its end.

The next day, on silky smooth currents growing wider, a seal bobbed up to look us in the eye—so human an animal with its curiosity and binocular stare, quietly announcing the upriver reach of tides.

We camped once more. The wind stopped. The river only whispered. The stars did their best to light the sky while a competing marine mist wafted inland, eventually enveloping us in fog. We watched in the dampening dark as the tide also crept toward our tent. Before turning in, we moved to higher ground and double-tied the raft.

On our last day, with ocean surf pounding on the horizon, Ann and I landed at Gold Beach. We packed our gear, and for a final good-bye we walked out on the jetty of rocks where saltwater sprays you in the face and where the current of the Rogue mixes with a forbidding swirl of breakers and ocean swells.

In the years that followed, we floated the Rogue many times, never tiring of its special flavor. We sought out the source at Boundary Springs, where the waters of Crater Lake—America's deepest—bubble up from dark volcanic passages. We hiked the Upper Rogue Trail past tumbling waterfalls and pools that reflect blazing colors of autumn and then enchanting whites of winter. Eventually we launched another "whole Rogue" voyage—this time covering 157 miles from Lost Creek Dam to the ocean, including two steep challenging drops on the middle Rogue upstream from Gold Hill. This great river's entire 216-mile route westward offers Oregon's best-known waterway passage through forests, mountains, and canyons.

Rogue River at Dulog Riffle
Rafting through the Rogue's canyon, which splits the towering Siskiyou Mountain Range, is one of America's most popular multi-day wild river trips.

National Creek Falls

At the wintry headwaters of the Rogue, National Creek's waterfall—reached by skiing several miles off Highway 230—freezes in a fortress of ice.

Middle Fork of the Rogue

Madrone trees rise over the Rogue's Middle Fork at the Route 992 bridge. Located at the crossroads of the Cascade, Siskiyou, and Coast Ranges—each with its own intricacies of geologic history and distinctive flora—the Rogue watershed is one of the most diverse in the West for plant life.

South Fork of the Rogue

Douglas fir, grand fir, and western red cedar blanket steep slopes above the Rogue's South Fork.

Bear Creek at Phoenix

Frost from valley fog breeds wintertime ghosts of willows and cottonwoods. A popular streamside bikeway follows this tributary through its cottonwood corridor from Ashland to Medford and beyond. Thanks to three dam removals on the Rogue River downstream, a healthy run of Chinook salmon, eager to spawn, returned to Bear Creek in 2014 after many decades when few fish could surmount the concrete gauntlet. In Ashland, it was the best salmon run in memory.

Illinois River below Buzzard's Roost

A quintessential American wild river, the Illinois offers one of the West's top expert whitewater voyages. Staggering left and right through a bewildering maze of wooded canyon lands, one-third of the river's length is sequestered in the Kalmiopsis Wilderness. Pristine tributaries call to native Chinook and steelhead, and the watershed supports most of the Rogue basin's threatened wild coho. A remarkable diversity of plant species bloom here in the Siskiyou Mountains' unusual soils. Local conservation groups work to reduce turbidity and habitat upheaval caused by suction-dredge gold mining, and also to halt ominous proposals by distant corporations for open-pit nickel mines. These would degrade tributary basins of the Illinois and of the California-bound North Fork Smith, along with Hunter Creek and Pistol River, which drain directly into the Pacific south of the Rogue.

Willamette River

Willamette River at Willamette Mission State Park

The largest river flowing entirely within Oregon, the Willamette presses south-to-north 187 miles and passes directly through the state's largest cities. Seventy percent of the state's population lives in this river basin.

If any stream qualifies as *the* river of Oregon, this is it. Providing crucial water supplies, wildlife habitat, and cottonwood corridors, it draws on fifteen major tributaries and hundreds of smaller streams, from the height of the Cascades' western slope to the eastern flank of the Coast Range with its shorter rivers stoked by winter rainstorms. The river offers one of the West's finest easy extended canoe outings, with camping on gravel bars and at parklands of the Willamette Greenway.

Willamette River downstream from Corvallis

Though shortened by the elimination of bends and back channels, riprapped to ditch-like simplicity compared to what once was, and farmed to the banks in many places, the Willamette is still lined by a green corridor of cottonwoods for most of its length. Even though it's thin, this riparian belt is most of what you see when traveling the length of this river.

Willamette River at Marshall Landing

A springtime rainstorm approaches on the Willamette downstream from the McKenzie confluence.

Willamette River and cottonwoods

Cottonwoods, willows, red osier dogwood, and wild rose begin to leaf out as springtime breaks along the upper Willamette.

Willamette Falls

Damming what was once among the most extraordinary natural features in the West, a barrier of concrete blocks the upper precipice of Willamette Falls. But when seen from below, the falls still thunders impressively. This photo was taken by canoeing upstream to the base of the falls and then scrambling partway up cliffs on the eastern side. Mist wafting from foaming water creates prisms for the low-angled sunrise and morning rainbow.

Canoeing the North Fork of the Middle Fork Willamette

This exceptional tributary to the upper Willamette begins at Waldo Lake—among the purest lakes anywhere. After dropping over waterfalls hidden from roads and trails, the North Fork riffles past an ancient forest at Constitution Grove and then sieves through the Miracle Mile of rocky whitewater. Just below, another section of rapids tumbles to meet the Middle Fork, which is an upriver extension of the main stem Willamette in all but name.

Middle Fork Willamette River below Dexter Dam

In sunset's glow, the Middle Fork Willamette riffles toward its confluence with the smaller Coast Fork. Together, just upstream from Eugene, they form the main stem Willamette.

McKenzie River below Sahalie Falls

With its source in Clear Lake—where you can still see underwater snags of trees inundated three thousand years ago when molten lava dammed the river—the McKenzie speeds downstream through old-growth conifers at Sahalie and Koosah Falls.

McKenzie River rapid

A popular footpath between Sahalie and Koosah Falls tours this river's extravaganza of enchanted woods and pummeling rapids. Continuing twenty-six miles, the McKenzie River Trail ranks among Oregon's premier riverfront hiking and mountain-biking routes.

McKenzie River with thimbleberry below Koosah Falls
Just below this green and flowered riparian scene, the McKenzie is dammed twice for hydropower, dating to a compromise that rejected dams and diversions at the showy waterfalls upstream.

McKenzie River at Koosah Falls
The McKenzie is the only stream west of the Cascade crest that has continually supported native bull trout—charismatic fish that have been reintroduced to the Middle Fork Willamette and upper Clackamas.

McKenzie at Tamolitch Pool

A full-bodied McKenzie, flowing five hundred cubic feet per second, incredibly disappears underground for three miles in lava tubes, then resurfaces from its unseen grottos in the sublime spring flow of Tamolitch Pool, reached by trail off Highway 126 upstream from Trail Bridge Campground.

McKenzie River above Trail Bridge

The upper McKenzie's steady hydrograph of spring flows allows moss to thicken on rocks down to the waterline.

North Santiam River

With snowmelt from Mount Jefferson—Oregon's second-highest peak—the North Santiam River floods into rich backwater pools and sloughs where tributaries join. This alder forest is at the mouth of Bugaboo Creek.

Santiam River at the North and South Santiam confluence

By far the largest Willamette tributary, the Santiam flows through a corridor of cottonwoods. Higher up, the North Santiam and South Santiam each course through Cascade gorges, rapids, reservoirs, and riparian forests. Both branches offer steelhead fishing and extended canoe trips that continue down the main stem to the Willamette.

Little North Santiam

This sizable undammed North Santiam tributary cascades from basalt cliffs and glimmers over black and white, marble-like rock here at Three Pools—mobbed by swimmers on summer weekends. Upstream, the headwaters of Opal Creek emerge from an exquisite ancient forest, secured as wilderness in 1996 after a decade-long campaign to halt logging there in Willamette National Forest.

Yamhill River at Dalton

The Yamhill eases through McMinnville before languidly joining the Willamette. The lower reach of this largest stream that flows from the east slope of Oregon's Coast Range was dredged for shipping and remains entrenched, but an entanglement of forest and shrubs once again covers the banks and shores.

Mollala River, Table Rock Fork
Largest among the few sizable Willamette River tributaries that remain undammed, the Mollala's headwaters reach to Table Rock Wilderness.

Table Rock Fork with big-leaf maples
Nested between the larger Clackamas and Santiam watersheds, the Mollala River drains lower Cascade Mountain terrain through deep gorges and rain forests of moss-clad big-leaf maples. It supports one of the best viable wild steelhead runs in the Willamette basin, and the lower river's cottonwood corridor has potential for restoration of salmon and wildlife habitat.

North Fork Silver Creek

North Falls carries the swelling runoff of autumn rainstorms in Silver Falls State Park while the trail arcs behind the 136-foot vertical drop. This Cascade foothills tributary later joins the Pudding River and the lower Willamette.

North Fork Silver Creek

On a rainy autumn day, the North Fork of Silver Creek plunges through Silver Falls State Park.

Clackamas River

The Clackamas bursts from Cascade Mountain slopes south of Mount Hood and foams down through a wonderland of forest and whitewater to the Willamette at the head of tides. Hiking trails follow part of the upper river's route, including a section here below Indian Henry Bridge. After a century of absence, bull trout have been reintroduced to Clackamas headwaters.

Clackamas River at Carter Bridge Rapid
Whitewater entices paddlers and rafters to a stairway of rollicking rapids through the middle Clackamas. Below here, three dams are followed by a riffling river at Portland's doorstep, featuring hiking, boating, fishing, and swimming at county parks.

Collawash River and cliffs
This largest tributary to the Clackamas flows from wild Cascade headwaters and past dramatic lava outcrops.

Collawash River above Hot Springs Fork
Formidable undercut boulders congest the flow of the Collawash.

Roaring River

One of the wildest rivers of the Cascade Mountains, the fourteen-mile-long Roaring has no roads alongside, but can easily be seen at a campground near the Clackamas confluence. The entire main stem and South Fork are protected as National Wild and Scenic Rivers.

The Columbia River and Its Gorge

With a timeless force stronger than rock, Oneonta Creek has pierced sheer walls of basalt and opened a crevasse-like canyon deep in the heart of the Columbia River Gorge.

Joining other weekend pilgrims, I parked along the Historic Columbia River Highway east of Multnomah Falls. I sealed my camera in a waterproof bag, slipped it into my daypack, and laced up my worst sneakers. Then I scrambled on a trampled path along Oneonta Creek to a formidable obstacle: basalt walls on both sides pinched the stream tightly enough to collect logs that fall from the cliffs above and flush downstream during high water.

Like a gate to secret worlds, the two-story logjam blocked not only my path up the creek, but also my view to whatever lay beyond. Yet the stream filtered through it all, and I could hear its distant rush on the other side, which doubled my desire to go.

Up to four feet in thickness, the logs were packed tightly enough that the entire logjam had the rigidity of a permanent structure but loosely enough that dark air spaces gurgled with unseen water beneath the husky boles. The spooky gaps were big enough to swallow me, or at least to snag a leg. So for traction I favored the trunks that hadn't had their bark stripped during their battering-ram journey downstream. I gangplanked the length of a full tree, grabbed a limb and pulled myself onto another log, then pivoted around an upright trunk that had been speared straight into the pile like a needle in a pincushion. Finally I eased myself down the other side, step by cautious step, to the level of the creek again.

With the logjam behind, I entered the sublime.

The walls of dripping black basalt ascended straight up. In some places their overhang blotted out the sky. Drops of

Oneonta Creek

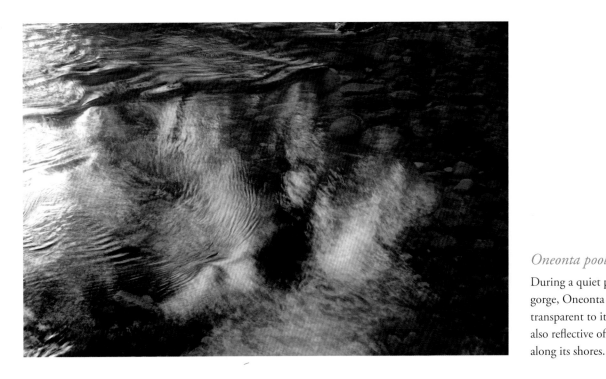

Oneonta pool

During a quiet pause in its gorge, Oneonta Creek is transparent to its depths but also reflective of autumn color along its shores.

morning dew fell from soft pillows of moss bedded on ancient conifers growing from the ledges and rimrock above. A few rays of sunlight angled into the hushed cathedral of shade.

In this world apart, with the roar of the interstate highway and transcontinental railroad left behind, and forgotten, I was surprised to find a diffuse parade of couples and small groups of young Oregonians wading up the creek, their happy voices subdued in the grandeur of the place.

As I walked, shallows deepened to my thighs, then ramped back up to dry gravel bars. But then the water grew much deeper. A young girl who wore only a bathing suit in the cool canyon gave up walking and swam. Two boys had opted to cling to the cliff, inching along with finger grips and toe holds that failed for one, dumping him harmlessly but with a shriek into the drink. Holding my camera bag over my head, I crept forward on slippery cobbles until the chilled stream water wetted my chest, stealing my breath. Then the bottom sloped up again to a final dry cobble bar.

At the end, where glistening fern-covered walls blocked every possible way forward, the white spout of a waterfall splashed to the canyon floor, and a few hardy young souls danced in the shower. This, too, was a river of Oregon.

Horsetail Falls

For seventy-two miles the Columbia River Gorge extends from the Deschutes River on the east side of the Cascade Mountains to the Sandy River at Portland's suburban edge. Following the distant breakup of Ice Age glacial dams upstream in today's Montana, incomparable floods carved the Gorge, leaving tributaries hanging high on walls that remained from earlier lava flows. The result is America's greatest concentration of waterfalls; seventy plummet from basalt cliffs on the Oregon side of the Columbia.

Columbia sunrise from Crown Point

The fourth-largest river in the United States delivers water from seven states and British Columbia. In Oregon, the Columbia's entire length above tidewater has been impounded by dams built for barging and hydropower. In the process, the world's most abundant salmon runs were pushed to the brink of extinction. In 1986 the Columbia Gorge National Scenic Area was legislated with the goal of protecting spectacular scenery of the corridor that's also shared with two dams, two railroads, an interstate highway, several towns, and a mosaic of private land that continues to face intensive pressures for development.

Eagle Creek waterfall

Eagle Creek tumbles down from high mountains and enters the heart of the Columbia Gorge. Its trail clings to cliffsides and offers the most scenic waterfront hike within a short drive of Portland. Footpaths continue to the upper basin and onward to Mount Hood's Timberline Lodge—a three-day backpack trip that ranks as the United States' ultimate hike from nearly sea level to alpine terrain.

West Fork Hood River at Punchbowl Road bridge

The Hood River's three forks collect snowmelt and glacial outflow from the north and east sides of Mount Hood and then stairstep down to the Columbia. A bridge north of Dee offers this unusual perspective of upright columns of basalt and cobbled braids of whitewater.

Sandy River above Zigzag

With winter snows lingering, the upper Sandy tumbles through glacial cobbles.

Sandy River and Mount Hood

This beloved stream east of Portland carries glacial runoff from Mount Hood to the tidal Columbia with the greatest undammed vertical drop of any river in Oregon—seventy-three hundred feet from the glaciers' outflow in alpine terrain to the level of ocean tides. Fish include winter and summer steelhead plus spring Chinook, fall coho, cutthroat, and rainbow trout. Upper reaches call to hikers, mid-sections challenge expert kayakers and rafters, and lower miles draw anglers, canoeists, drift boaters, and anybody in the Portland area who's inclined to spend an afternoon along an agreeable river.

Salmon River above Old Salmon River Trailhead

A prize of Oregon's north Cascades, the Salmon River flows from the south face of Mount Hood to the Sandy River. Ancient forests, enticing paths, and whitewater riddled with logjams can all be reached by trail from the Salmon River Road.

Sand Canyon Creek above Zigzag River

Shaded by hemlock, grand fir, and western red cedar, this tributary to the Zigzag and Sandy Rivers arcs around a sharp bend on the lower slopes of Mount Hood.

Deschutes River

Not counting the Columbia and Snake, whose flows come mostly from other states, the largest river in the vast area east of the Cascade crest is the Deschutes. Next to Hells Canyon of the Snake, this is the premier big-water river trip in Oregon, with a summertime push of five thousand cubic feet per second and the feel of raw power in its breaking waves.

For the longest river expedition possible, avoiding dams blocking the Deschutes' path, Ann and I launched our raft and kayak at Warm Springs above the Highway 26 bridge. From there the river stretches ninety-eight dam-free miles to the Columbia, with some of the West's most famous trout water and also thrilling rapids in an impressive desert canyon.

The Deschutes runs in marked contrast to intimate forest-shaded streams like the coastal Siuslaw, which spike immediately when winter rains pummel their local basins. Here, the lower river is all desert, with flows emanating not from the hills you can see, but from remote landscapes and hidden Cascade groundwater as far as two hundred miles away. Mountain snowmelt filters through lava tubes and percolates in subterranean water tables to yield one of the West's steadiest hydrographs in downstream reaches, which peak at only 1.7 times the low flow (some rivers peak at forty times their smallest discharge).

While rivers of western Oregon flow through impenetrably thick forests, here east of the Cascades we searched for single trees and the shade they offered. In that sere landscape, water is all the more precious for its scarcity. The shrubby green riparian edge, the host of insects buzzing low on the food chain, the fish jumping, birds singing, and furbearers slipping in and out of the flow all depend on the river's thin strip of wetness and on runoff from distant mountaintops.

Ann and I were a bit giddy with the simple joy of floating on such glassy water, peering into see-through depths as we drifted the first ten miles. Much of our route northward was shared by a railroad, but we didn't pay it much attention, and highway crossings were pleasantly scarce during our week-long sojourn.

On the second night we camped at Davidson Flat, where a tunnel swallowed the train tracks, leaving us alone on a peaceful bowknot bend. Sweeping us downriver the next two days, currents accelerated, and we eddied out to scout several large rapids with holes and rocks to be avoided.

About halfway through our journey we beached at the mouth of the White River. Unlike most tributaries, which burst clean and steady from volcanic springs, the White delivers milky white flows pureed from rocks that grind against each other as the Mount Hood glaciers advance downslope. Unlike other Deschutes shores typified by blackened volcanic basalt, springtime freshets here had dropped a whole plain of glacier-rounded cobbles and buried the lava bedrock.

In a few more bends, at Sandy Beach Access, we met a shuttle driver for a truck-and-trailer portage around Sherars Falls—an unrunnable drop of fifteen feet—with added gradient in a preface of accelerating whitewater and then, below, with a whirlpool epilogue. Salmon migrating upriver encounter a foaming bottleneck here that forces them to jump in predictably vulnerable places. Knowing this, as their ancestors have for the ages, Indians stood on rickety overhanging platforms guy-wired to the canyon

Deschutes River at Trestle Hole Island

walls so they could dip net for the jumping fish, much the way they once did on the Columbia at Celilo Falls before it was terminally flooded by backwater of The Dalles Dam. Here on the Deschutes, Sherars is the closest living replica of Celilo's legendary nexus of salmon, waterfall, and tribes' dependence on the life of Oregon rivers.

Downstream, the orange and brown walls of basalt rose higher, and our route passed Macks Canyon as we entered the last twenty-four-mile roadless reach of the Deschutes.

Two days later, we splashed through the elegant Gordon Ridge and the rowdy Colorado Rapids, and in four more miles eased to our takeout at the lap of the reservoir pushed upstream by The Dalles Dam on the Columbia.

Just as the Siuslaw had toured the quintessential coastal rain forest and the Willamette had showcased Oregon's central valley, the Deschutes revealed to us the potent vitality of a river flowing through desert canyons on the east side of the Cascades.

Deschutes River at Davidson Flat

Basalt walls veer up from the Deschutes in its deepening canyon.

Deschutes River above Wickiup Reservoir

This gauntlet of logjams appears in the short, rambunctious interlude between the upper Deschutes' two Cascade Mountain reservoirs. Water storage for agriculture acutely lowers the river below those dams in winter, causing fish kills. With higher flows released in summer, the Deschutes is a recreational hotspot through Sunriver and Bend where trails and bikeways wind along miles of rapids and quiet pools—one of the finest urban greenways in the West. Downstream, the middle river is depleted by farm diversions in summer. Finally, below Pelton Dam, flows are restored to the lower river.

Paulina Creek Falls

Below its source at the water-filled crater of Paulina Lake, southeast of Bend, this feisty creek careens over a double waterfall in Newberry National Volcanic Monument.

Paulina Creek

Paulina Creek free-falls from its crumbling volcanic precipice and downward toward the Little Deschutes River.

Whychus Creek

Windblown spray of Chush Falls has veneered trees, logs, and rocks in ice on a cold autumn morning. Whychus Creek begins on the glacial slopes of the South Sister and flows to the Deschutes River.

Metolius above Gorge Campground

The Metolius's first few miles coast swiftly with no large rapids. Then below Gorge Campground, two steep pitches are riddled with waves and holes. Finally, through a lower canyon, relentless swift current brims against both banks and sieves through dense brush and fallen logs. The river ends in flat water impounded by Warm Springs Dam on the Deschutes.

Metolius River below Camp Sherman

The quick, frigid, crystalline water of the Metolius is renowned for trout fishing northwest of Sisters. The river rises fully formed out of a spring bubbling forty-five thousand gallons per minute within sight of Mount Jefferson's pointed snowcap. Here is one of Oregon's few robust refuges for bull trout, which migrate the length of the stream.

Crooked River at Smith Rock State Park

The Crooked and its South Fork headwaters roam for 203 miles, almost as far as the better-known Rogue. But owing to the Cascade Mountains' effective rain shadow, flows here are lean. Productive trout waters riffle for eight miles below the chilled release of Bowman Dam. Then diversions deplete the small river. Here at Smith Rock State Park, the stream crosscuts though a monolithic plug of volcanic rhyolite.

White River and Mount Hood

The White River races with runoff from the eastern flanks of Oregon's most iconic mountain and down to the Deschutes. Glaciers clinging to Hood distinguish this stream with an icy backdrop and floodplain cobble bars upstream from Highway 35. Midsections hide in wooded canyons, while Tygh Valley State Park features a spectacular chain of waterfalls near the mouth, marred by abandoned hydroelectric equipment.

Eastern Oregon Rivers

Scenic highlight of the enormous John Day basin, and its key refuge for wild Chinook and steelhead, the North Fork carries twice the volume of the main stem where the two meet in the crossroads of Kimberly. Together they make the West's longest river voyage of top quality that's also relatively easy to negotiate—225 miles from Highway 395 on the North Fork to McDonald Crossing near the Columbia.

Like the neighboring Deschutes, the John Day starts in mountains and then winds through Oregon's desert. But without the saturated upstream sponge of groundwater and the lingering snowdrifts of Cascade peaks, it drops quickly after snow melts from the Blue Mountains. The John Day also lacks the Deschutes' midway dams, diversions, and waterfalls. Instead, it navigates the entire north-central sweep of Oregon's map without major interruptions to river travel in springtime.

Drawn to this geography, Ann and I set out on our John Day rafting adventure at the end of May. As we curved down to the North Fork put-in, a late-season squall on the highway summit to the south mellowed to a chilly breeze. But then the weather warmed daily with the exhilarating breath of spring in the mountains. Soft to the skin, mild morning breezes ripened with the glow of sunshine, which thoroughly thawed all memories of winter's persistent snow and rain.

We found mile after mile of swift, clear water, a few lively rapids, and ridgelines that begged us to stroll for breathtaking views during quiet evenings or at daybreak as the sun eased up with yellow rays. The place perfectly fit my expectations of paradise.

The landscape began to dry out even during the first day of travel as a thick forest of pines and Douglas firs yielded to pines only, then to a splendid pine-and-grassland mosaic where wildfire maintained savannas for deer and chattering woodpeckers in their dipping flights, tree-to-tree. The drumming birds fed on dead snags and kept the pine-bark beetles in check. Meanwhile, rapids danced over basalt bedrock. Campsites awaited on gravel bars or shaded banks. We found ourselves wanting to stop and stay at every bend, but pressed onward with miles to go on our two-week schedule.

The North Fork and main stem John Day combination is Oregon's longest river flowing entirely within the state (though the Willamette is far larger in volume). It's also the longest dam-free section of river in the entire Northwest—284 miles from the source, above our put-in, to almost river's end in the reservoir formed by John Day Dam on the Columbia. Free-flowing mileage, along with a near-absence of development, are key reasons why the North Fork boasts Oregon's best remaining salmon runs above Bonneville Dam. The John Day is also unaffected by hatchery fish, which elsewhere in the Columbia basin and much of Oregon compromise genetic integrity of native stocks.

As we dipped our oars into swift currents, the miles sped by. Above the village of Monument, ranchlands took over the shorelines for a while, and then at Kimberly the main stem and its comparatively small volume nudged us onward. A day later, at Service Creek, we entered the larger canyons that carve deep into golden rock with intermittent ranchland for 148 more miles. Riffles and easy rapids intensify just once, at Clarno, where canoes are portaged but rafts splash through the big waves.

Along the lower river, with our eyes peeled for bighorn sheep and golden eagles, we were awed by rock walls of

North Fork John Day upstream from the Middle Fork confluence

a thousand feet. At the lower end of the John Day, the Western Rivers Conservancy bought eight thousand acres of Cottonwood Canyon in 2008. This became Oregon's largest state park, managed as a wild preserve.

Nearby, at the Cottonwood access, our adventure that had begun on a blustery spring day in the forested Blue Mountains ended where the rocks were hot to the touch and the river eased into its final thirty miles leading to the backwaters of John Day Dam.

John Day River aerial view

In winter, the John Day twists northward from Clarno—at the top of the photo—and past the impressive uplift of Iron Mountain on the left in this view from thirty thousand feet.

John Day River below Service Creek

A steppe of grassland, morphing from green to gold with the drought of summer, ramps up to ridgelines of the Columbia Plateau.

Middle Fork John Day above the North Fork confluence

Downstream from Highway 395, the Middle Fork John Day charts nine roadless miles through pines and grasslands to its mouth. I came to this springtime scene beneath ominously brewing skies while hiking upstream from the North Fork confluence.

Middle Fork John Day west of Highway 7

Upper reaches of the Middle Fork John Day riffle past meadows and forests of ponderosa pine and Douglas fir. The stream's quiet paved byway makes for excellent riverfront bicycling in the spring when hills turn from brown to brilliant green.

Klamath River below Boyle Dam

Beginning in south-central Oregon and then curving an epic route across northern California, the Klamath becomes the third-largest river entering the Pacific this side of Canada. Downstream from hydropower dams in the Klamath Falls area, the flow thunders through a steep, wild, wooded canyon to backwaters of dams near the California border. This is one of only three rivers that begin east of the Cascade Range and cross both it and the Coast Range (the Columbia and California's Pit/Sacramento are the others, passing through wider gaps). Algae from warm, nutrient-rich, upstream reservoirs tint the water sickly green and paint black slime on rocks exposed at low flows in summer, seen outlining the shores here. For a hundred miles downstream of this spot, algae accumulate to toxic levels, plaguing the Klamath and its salmon and steelhead.

Klamath and Oregon white oak below Boyle Dam
Autumn colors brighten the Klamath's canyon through the Cascade Mountains. A pending plan for restoration would eliminate four dams and open the upper river to spawning salmon, though elevated water temperatures from the farmed and cutover upper basin will remain troublesome without further restoration.

Chewaucan River

A welcome surprise in the vast landlocked drylands of southcentral Oregon, the Chewaucan (she-wau-can) headwaters drift from pine-clad mountains that capture precious snowfall. The lower river is heavily diverted for ranches, though in recent years restoration efforts by the Department of Fish and Wildlife and landowners have reclaimed promising trout habitat. Following the pattern of ancient glacial rivers that once nourished immense Pleistocene lakes, the Chewaucan still supplies Lake Abert—eastern Oregon's only landlocked lake that has not completely dried up in historic times.

The Snake River

With runoff beginning in Yellowstone National Park—650 miles upstream of Oregon—and then crossing western Wyoming and southern Idaho, the Snake River ultimately becomes the twelfth-largest river in the United States and by far the largest tributary to the Columbia. Defining the Idaho-Oregon boundary, Hells Canyon offers Oregon's premier big-water raft trip.

The high-volume flow and boiling hydraulics here can overturn boats of any size if they end up in the wrong place, so I asked Bill Sedivy—an old friend and director of the conservation group Idaho Rivers United—to join me and Ann as we launched our Snake River journey in mid-July. We would ride on seventeen-thousand cubic feet per second released from Hells Canyon Dam, located just above our put-in. I had done this trip only once—twenty-six years before—so took comfort in Bill's company.

Though the 110-degree heat might have justified the name "Hells Canyon," the scenery did not. I admired countless layers of volcanic and colored metamorphic rocks that piled up on top of each other to grassy benches, scattered ponderosa pines, forests of fir, and finally ridgetops looming far overhead, altogether a topographic extravaganza. The canyon's full enormity is never seen because canyon walls and sub-peaks block the view to even greater heights. It's a museum of biological diversity spanning a mile and a half of vertical rise above continental cleavage separating the Seven Devils Mountains of Idaho from the Wallowa Mountains of Oregon. In the United States, only the Kings River in the Sierra Nevada carves a deeper canyon, though the Snake is much larger and flows through a longer canyon corridor.

After an hour of floating and acclimatizing to the aggressive flow, we came to the first big rapid, Wild Sheep. I couldn't yet see it, but given the scale of the place, I knew it was going to be enormous. From our upstream, river-level perspective, the flow thundered over a horizon-line agitated with splash, then disappeared. We eddied left for a careful scout from shore.

Perched on a rock, Bill pointed to the nonnegotiable route. Gripping the oars again, I rowed into the current, entered on a tongue of green mint jelly, pulled right as the glassy surface disintegrated into foam that scoured jagged boulders, then pulled harder to avoid the mayhem of whitewater on rocks below. A wave suited for a rough day on the ocean welled up and then blasted the left side of the raft, punching it skyward. With instincts refined by years of rafting experience, Ann expressly shifted her weight to that rising tube, "high-siding" to hold it down. Then we were through, and we laughed in the joyous excitement of it all.

Through one of the biggest rapids in Oregon, the unleashed power of nature had given us a rare thrill that combined fear and exuberance, surrender and triumph. The river had taken us on one of life's memorable rides and, having survived, we were left with a new feeling of belonging there at the heart of a force that's utterly fundamental to the workings of the earth: gravity pulling water downhill, the same way it does on every river of Oregon, but with shameless drama here.

In another two miles, Ann rowed through Granite Rapid's similarly riotous hydraulic path. In three more miles we eddied out to camp in the welcome shade of alder trees at the mouth of Saddle Creek.

All three of us cherished the next two days. Could life be any better? Then, with the largest rapids behind us, Bill got out at Pittsburgh Landing, where a shuttle driver had delivered his truck by way of a road that switchbacks up and

Snake River in Hells Canyon
Layers of volcanic and metamorphic rocks rise up above the Snake River downstream from Hells Canyon Dam.

out of the canyon on the Idaho side. Ann and I continued for another three days of delightful transit on wave trains, past impressive views of harsh "rockscapes" rising in golden light, and to the heavenly mouth of the Imnaha River bubbling with refreshment. Each scenic episode built on the geographic story that this river revealed to us hour by hour, day by day.

Then we came to the mouth of the Salmon River. Draining all of central Idaho, it starts in the venerated Sawtooth Mountains and brushes by seven major subranges of the northern Rockies, bisects the largest wilderness in the West, and provides the finest habitat for Columbia basin salmon. It increases the Snake's volume by another third.

Just below the confluence, a proposed dam would have flooded both the Salmon and Snake. A lot of bad dams have been built in the United States, but this would arguably have been the worst. That plan was dropped, but just above the confluence an alternative called High Mountain Sheep Dam was plotted. There, after a classic battle in the history of river conservation, the 1975 Wild and Scenic River designation for the Snake stopped the dam. The Hells Canyon battle turned the tide of dam building.

A few more large dams were constructed, but within a decade the era of new mega-dams in the United States (but not the world) would be over. At least for now.

Drifting through the calm waters where the seventy-story skyscraper of concrete had been proposed, we felt grateful that people had stood up courageously when they were needed to protect the nature of this place in all its native glory rather than flooding it with one more flat-water reservoir.

As we touched ashore at the Heller Bar takeout at the end of our six-day trip, Ann and I couldn't help but regard this thrilling ride through the Northwest's greatest canyon as a celebration of many people's collective efforts to protect wild places, and as a perfect finale to our tour of Oregon rivers.

Snake River at the mouth of Saddle Creek

Snake River at sunset, Sulphur Creek Rapids

Flows of the Snake River are essential to salmon that migrate nine hundred miles to spawning grounds in Idaho's Salmon River basin. Once numbering sixteen million or more in the Columbia basin, these muscular fish have been reduced to endangered species by dams on the lower Snake. The Save Our Wild Salmon Coalition wages a campaign to eliminate the four dams that block the Snake below Lewiston, contending that they provide only 4 percent of the Northwest's electric power, that transport of commodities by heavily subsidized barges is more economically done by existing rail routes, and that no other solution is possible to avoid further decline and likely extinction of wild salmon.

Owyhee River upstream from Rome

The Owhyee is one of the more difficult-to-reach rivers in Oregon. A lone paved road crosses at Rome, separating two great canyons, upstream and down. Boating the upper canyon requires driving on rough dirt roads during the narrow window after winter's lingering freeze but before snowmelt withers.

Owyhee River canyon upstream from Rome

One of America's quintessential desert-canyon rivers, the Owyhee runs for 249 miles through the dry lands of Nevada and Idaho. Then, near southeastern Oregon's border, three branches converge and the river engraves a path through vertical-wall canyons for another ninety-six nearly roadless and undeveloped miles to Owyhee Reservoir.

Upper Malheur River

Most travelers along Highway 20 through eastern Oregon know the Malheur only as a diverted, parched, weedy remnant of what once was. But, along with its North Fork, the upper main stem flows from wild enclaves of the Blue Mountains through Edenic pine savannas and unexpected groves of arrow-straight larch trees fattened to five feet thick. A National Recreation Trail parallels the waterfront and bull trout lurk in deep pools.

Pine Creek east of Halfway

After tumbling from granite peaks of the Wallowa Mountains, Pine Creek irrigates ranches and then wanders through cottonwoods at the base of grassy slopes and finally to the slackened backwaters of Hells Canyon Dam.

Imnaha River below Cow Creek Bridge

Secluded in northeastern Oregon, the Imnaha connects a wonderland of Wallowa Mountain high country and conifers to cottonwood floodplains and deepening canyons—a whitewater gem within a mile-deep valley.

Imnaha River

Flowing into Hells Canyon, the Imnaha is the Snake River's uppermost major tributary still accessible to salmon. It's also home to bull trout and native rainbows.

Grande Ronde River at Sheep Creek Rapids

This lifeline of northeastern Oregon winds through woodlands and then into dry canyons before it enters Washington and joins the Snake. With dazzling green forests of early summer, the river above Troy offers good fly-fishing and has adequate flows for boating well into summer and in some years through autumn.

Grande Ronde below Alder Creek

Beginning at the Highway 82 crossing in Minam, rafters, drift boaters, and canoeists ply the delightful lower Wallowa River to its confluence with the Grande Ronde and onward to Troy for a forty-seven-mile trip.

Grande Ronde River at Horseshoe Bend

Downstream of Troy, a paved road follows the Grande Ronde—a rare, seldom-run eastern Oregon river that can be paddled all summer—with wave trains passing ranches, pines, and grassland. Then in Washington, a roadless and increasingly arid twenty-six miles climax with the turbulent Narrows, followed by Bridge Rapids before the Grande Ronde swirls into the Snake River.

West Fork Wallowa River, lower gorge

Above Wallowa Lake—where vacationers flock to exhilarating high country that rivals the finest glaciated ranges of the Rockies—the Wallowa's East and West Forks merge. Wandering off the main trails, I found this hidden gorge upstream from the river's Wallowa Lake inlet.

East Lostine River and snow bridge

Like the nearby upper Wallowa, the Lostine and East Lostine flow from alpine meadows and peaks where snowdrifts linger into summer. A sixteen-mile loop trail tours both these magnificent streams. Just to the west, the Minam River embarks on an even wilder course to the Wallowa River.

Wenaha River

Longest completely wild river in Oregon, the Wenaha courses through the Wenaha-Tucannon Wilderness west of Troy for almost all its mileage before meeting the Grande Ronde. Bull trout heaven, these cold waters tour remote canyons and picturesque ponderosa savannas reached only by trail. The Wenaha's entire length is designated a National Wild and Scenic River.

East Lostine River and Eagle Cap

The trail up the East Lostine skirts meadows and wetlands with mesmerizing views to headwaters on Eagle Cap's 9,573-foot summit, glowing here during a July sunset. The Lostine and five other streams—Minam, Powder, Eagle, Imnaha, and Wallowa—radiate spoke-like off the Wallowa Mountains' uplift. Trails along these streams offer Oregon's finest streamside hiking in high country.

Drift Creek
This exceptional coastal stream riffles through the Drift Creek Wilderness before joining the Alsea River.

CHAPTER 4 Rivers Through Time

Deep within every river lies a story about the changes that the stream has undergone through time. During the two-week trip down the Willamette River described in Chapter 1, I saw the river of today, but also patterns of the recent and distant past. As I continued my photo expeditions across Oregon, I saw those patterns almost everywhere.

Today's typical view of the Willamette, which is mostly of riffling current and cottonwood forests, looks like you might think it has always looked. In many places the eyeful of green and blue, in fact, does reflect a distant past when the Willamette ran wild and free with no dams on tributaries, no logging on mountainsides, no farming on floodplains, and no cities at all. But a closer look reveals many changes that have diminished the river in one way or another. Rocks and even old cars have been dumped on the eroding outsides of bends. The cottonwood corridor—impressive and grand at first blush—is often just one or two trees deep, while in the past those trees ramped back for miles with multiple sloughs nourishing wetlands rich in life. It was not just a green riverfront, but a vast and far-reaching ecosystem. Where there used to be splashing crowds of salmon and darting native cutthroat, sculpins, and whitefish, we now have scavengers such as catfish and carp and exotic predators, including bass and German brown trout.

Yet ongoing recovery from some of those setbacks is also evident. One day I postponed stopping for lunch to avoid the stink of sewage below one of the larger towns, but the water for the entire length of the Willamette was remarkably cleaner than what I would have encountered fifty years ago. Sediments in the lower river remain saturated with residue of hazardous wastes from paper mills, but just a few decades back those or other poisons might have been toxic to the touch.

The Willamette's complex transition signifies the story of Oregon's rivers everywhere. It's a tale of discovery, development, and loss that has spanned a century and a half, and it's increasingly a story of appreciation, protection, and restoration. The story of the Willamette unmasks a remarkable evolution of our outlook: from ignoring the intrinsic value of streams to cherishing the beauty and life they support.

Yet conflicts continue. Furthermore, the projected growth of population in Oregon means that demands on our water and rivers will increase, and keep increasing, until we act on the belief that unlimited growth is a liability, not an asset, to the quality of our lives and those of generations to come. The way that Oregonians address these threats, trends, and possibilities will say much about the fate of our waterways, our communities, and ourselves. In this way, the rivers reflect much more than the shorelines alongside or the skies above.

The histories of Oregon's development and of river degradation are inextricably linked. Consider these numbers:

Thirteen hundred large dams, over ten feet high, block almost every major river.

Fifty-five thousand diversions, according to the Department of Water Resources, take water out of rivers for industries, farms, and homes.

Thirty-five species of fish are threatened or endangered under state law; sixteen are federally listed.

Salmon have been reduced to less than 5 percent of their original numbers.

Bull trout have been eliminated from 96 percent of their former range.

Fourteen thousand miles—26 percent—of surveyed streams have serious water quality problems, according to the Department of Environmental Quality.

Seventy-one percent of stream miles are degraded when the added criteria of water temperature and biological diversity are considered, according to the *Oregon State of the Environment Report*.

Responding to these problems, Oregonians have rallied with campaigns to protect and restore their rivers. Each new initiative has built on the one before it. Those citizen efforts began with concerns for the quality of the water we drink.

In the first wave of river conservation, people sought to curb the most flagrant dumping of sewage from cities and poisons from industries, which posed virulent hazards to public health. In 1938, after a bill to address the problems was vetoed by Oregon's conservative governor, by a three-to-one margin voters elected to create the Oregon Sanitary Authority. Despite this remarkable show of support even in the depths of the Depression, regulations lacked teeth, the legislature withheld funds, and political resolve floundered under the grip of the wood products industry and its fear of regulation.

A turning point came when newscaster Tom McCall produced *Pollution in Paradise*—a 1962 television documentary stinging with investigative exposure. Springing from this notoriety, the charismatic McCall launched his political career. He won a successful bid for governor in 1966.

Cleaning up the Willamette became one of McCall's signature programs, and an era of river restoration began, including plans for a Willamette Greenway intended to link together public open spaces along Oregon's heartland waterway. Governor Robert Straub continued and improved upon McCall's conservation initiatives. Water quality standards for swimming were met in the 1970s, and fish began to recover. Statewide, cities and industries along other rivers followed suit.

After the most blatant problems had been addressed, the Department of Environmental Quality reported that the worst pollution remaining was runoff from large areas of disturbed land. Along the Willamette River, 80 percent of the pollution comes from farm drainage—also troublesome in the Tillamook, Klamath, Malheur, Long Tom, and Pudding basins. Though even today agriculture remains largely exempt from pollution laws, farming causes silted runoff that's often laden with pesticides, feedlot waste, and the nitrogen overload of chemical fertilizers.

Voluntary solutions have been adopted by some farmers, including reduced use of herbicides, the routing of runoff around feedlots, fencing to strategically direct cattle's access to stream fronts, water conservation, and organic farming. More of all this needs to be done, and larger buffer strips of native vegetation along rivers are needed to intercept runoff, filter pollution before it reaches the shores, and shade the streams.

Vast acreage in Oregon has also been affected by clear-cut logging that strips the soil of plants and then triggers erosion. This well-documented problem became clear to me one autumn morning along the lower Umpqua. After a rainstorm I walked through a clear-cut area where most of the dirt was exposed, and I noticed a single alder leaf strangely perched on an inch-high, vertical-sided pedestal of soil. All the dirt around it had been washed away by the pelting raindrops of the latest coastal storm. For the moment, that single leaf had remarkably protected the soil beneath it from erosion. Imagine how effective a full foot or two of the undisturbed forest floor's spongelike soil

South Fork Coquille

This landslide has occurred at the base of a massive and recent industrial timber harvest. Virtually the entire "buffer strip" has slid into the river.

would have been, with its enrichment of needles, leaves, mosses, fungi, and organic mulch. Instead, the exposed soil erodes and oozes into streams where it clouds the water in levels abrasively lethal to fish, smothers spawning beds, and chokes entire channels with muddy debris.

Even worse, haul roads for logging are scraped out by bulldozing through the virtual bio-membrane of roots and organic matter that makes up the top few feet of forest soil. Like a knife cut through skin, these roads bleed the flow of water that would otherwise seep slowly beneath the surface, nourishing the entire body of the forest and resupplying its groundwater. Ditches dug along logging roads carry runoff when it rains, and when it rains hard, torrents erode the ditches deeply, destabilizing whole mountain slopes and triggering landslides. Pacific Watershed Associates reported that 71 percent of slides inventoried after Northwestern storms in 1996 originated on recently logged slopes. To reduce damage, improved forest practices could increase setbacks from streams, spare the steepest slopes where problems are almost guaranteed, and minimize road construction.

The Oregon Forest Practices Act offers some protections in this regard, but not enough to avoid erosion, landslides, and stream pollution, according to scientists Reeves, Burnett, and Gregory in *Forest and Stream Management in the Oregon Coast Range* (2002). The Environmental Protection Agency and National Oceanic and Atmospheric Administration reaffirmed this view with a ruling in 2015 that the state fails to protect fish and water from pollution caused by clear-cutting next to streams, by runoff from logging roads, by logging-induced landslides, and by aerial spraying of pesticides. Even within prescribed stream setbacks limited to one hundred feet, the state's rules allow large trees to be cut down to a sixty-foot setback line. No setbacks are required if fish aren't found in the water.

I saw the law's shortcomings while canoeing on the South Fork Coquille below Baker Creek in 2012. Just after a massive industrial clear-cut was logged on slopes above the shores, the entire hundred-foot-wide "buffer strip" slid into the river, pouring silt onto premier spawning beds of salmon downstream. When I questioned state officials, they maintained that the requirements of the law had been met. The question remained: How can those requirements be considered adequate?

State regulations in 2015 also allowed aerial herbicide spraying only sixty feet from fish-bearing streams—as if the wind didn't blow. Direct spraying was permitted on streams too small for fish—as if small streams didn't harbor other life and immediately run into larger streams. Enforcement of even these regulations was lax, as residents of Cedar Valley near the lower Rogue found when their houses, gardens, and domestic animals were directly sprayed by a helicopter applying toxic herbicides to industrial forest lands nearby, sickening both animals and people, as reported in the *Oregonian* on October 22, 2014.

A big stride in timber management came with the Northwest Forest Plan of 1994, though it addressed only federal tracts mostly west of the Cascade crest. The policy requires broader stream-front buffers and no-cut zones within habitat of imperiled species such as marbled murrelets. The landmark reform remains under fire from those who want to log with fewer restrictions.

Pollution problems can be reversed—given adequate political will—but rivers literally disappear when dams permanently flood them with reservoir flat water. In the process, the streams' intricate flows, riparian habitat, valleys with farms and homesteads, and canyons of wild splendor are all sacrificed. Dams block the migration of salmon, steelhead, bull trout, and sturgeon, and often benefit alien fish. They diminish flows downstream when the water is held back in reservoirs or diverted for hydropower or irrigation. Many reservoirs increase water temperatures. By muting the effects of floods, dams transform the shape of riverbeds far downstream, eliminating pool and riffle sequences and the migration of gravel essential for river health.

In one of the nation's earliest actions to spare a river from damming, Rogue River sportsmen in 1921 persuaded the legislature to prohibit new dams downstream from Butte Creek. But that was an exceptional case during a half-century frenzy of dam building. At the Columbia River in the 1930s, biologists presciently forecast the decline of salmon due to the hydropower dams being built; however, the message was lost in the Depression-era din of dam-boosters. Back-to-back Columbia reservoirs ultimately reduced some of the world's greatest runs of salmon to the point that some populations were endangered and some became extinct.

Fighting those trends, McKenzie River enthusiasts rallied to defeat an Army Corps of Engineers' dam proposal at the mouth of Quartz Creek in 1947. Their bittersweet compromise allowed tributary dams to be built on the South Fork and Blue River. Additional McKenzie dams plotted at the spectacular Koosah and Sahalie waterfalls by the Eugene Water and Electric Board were rejected in a 1956 vote by Eugene residents. But the Board simply moved its project downstream to avoid the choicest scenery.

Trade-offs like these, allowing dams on alternative sections of rivers, typified early protection efforts but ended with Hells Canyon of the Snake River. Idaho Power Company had gained approval for three dams that impounded ninety-four miles in the upper half of Hells Canyon—a place that veteran river runner Martin Litton considered comparable to the Grand Canyon for wildness and rapids. Then, downstream in the 1960s, even larger plans were drawn by the federal Bureau of Reclamation for a seven-hundred-foot dam named Nez Perce after the tribe that had been evicted from its cherished homeland.

The dam would have flooded what remained of lower Hells Canyon and also Idaho's Salmon River—prime spawning ground for Columbia basin salmon and centerpiece of the greatest wilderness area left in the West. Nez Perce Dam was stopped, but the unwelcome alternative was the 670-foot High Mountain Sheep Dam on the Snake River just *above* the Salmon River confluence. Lawsuits involving competing private versus public dam sponsors, which both wanted to build this edifice, escalated to the Supreme Court. In a landmark ruling, Justice William O. Douglas directed that the Federal Power Commission consider the public benefits of building *no dam at all*.

With little attention to this requirement, the plans proceeded, and only minutes before appeals were due, Sierra Club lawyer Brock Evans filed objections. The resulting delay allowed time for Northwestern conservationists to build a national campaign, and in 1975 the remaining half of Hells Canyon was added to the Wild and Scenic Rivers system, banning further damming.

In another pivotal action in the 1970s, citizens blocked a proposed dam on the South Santiam that would have flooded Cascadia State Park. They revealed that the Army Corps had ludicrously inflated the project's value by averaging flood control benefits of all dams in the Willamette basin. Likewise, a South Umpqua dam near Days Creek was dropped after the General Accounting Office exposed its economic analysis as a sham.

These cases in the 1970s marked a turning point in the support of large dams in Oregon, but like a supertanker, the momentum to build lumbered onward and required a long time to stop. In the Rogue basin, residents sponsored an initiative to halt Applegate Dam, arguing that 80 percent of the "benefits" were for downstream real-estate speculators and that fifty miles of salmon streams would be blocked. Pro-dam boosters rented billboards and claimed that the measure would sacrifice 160,000 jobs. That many jobs didn't exist in all of southern Oregon, and none depended on the Army Corps project, but the vote to save the river was lost, and Applegate was built.

Finally ending the era of big-dam construction in Oregon, the Corps' Elk Creek Dam in the Cascade foothills met resistance from conservation groups because it would flood a third of the spawning habitat of the Rogue basin's imperiled coho salmon. The 186-foot-high cement wall was halted one-third of the way through construction by an Oregon Natural Resources Council (now Oregon Wild) lawsuit in 1987—a turnaround scarcely thinkable a decade earlier.

With these battles, the river conservation movement in Oregon gained stature and shifted into restoration mode. Proposals to dismantle Elk Creek Dam eventually garnered even Army Corps support, and the unfinished concrete plug was notched down to river level in 2008, allowing the coho to return. During the next three years, dam removals on the Rogue at Gold Hill, Savage Rapids, and Gold Ray allowed salmon, steelhead, and cutthroat to spawn; the Rogue became the nation's foremost example of a reclaimed stream. The fifty-foot Marmot Dam on the Sandy was also removed. Small dam removals on the Calapooia, Sprague, and Hood Rivers opened miles of additional streams to migrating fish.

Larger in scale and engrained in the political culture of river development, four dams on the lower Snake River, each a hundred feet high, fatally hinder salmon and steelhead on their journeys up and down river. The dams are in Washington but arrest fish bound for Oregon and Idaho. Numbers of wild salmon even in rare good years totaled far below thresholds for the species' long-term survival, and the Corps' solution—trapping young fish and motoring them in barges to sea level—failed to recover the threatened populations. Spilling more water over the dams during migration season has helped in recent years, as have cyclical ocean conditions, but pressure to maximize hydropower by reducing the spill has persisted.

The four fish-killing dams generate less than 4 percent of the Northwest's electricity—mostly in springtime when it's least needed. The power could easily be replaced through efficiency measures, according to the Northwest Power and Conservation Council. The impetus for building the dams had been to motor barges full of wood products and wheat from Lewiston—450 miles from the ocean. But that barging system, with commodities now bound for Asia, is 91 percent subsidized, and only a small fraction of the Columbia River's traffic motors the whole way from Lewiston. If the four dams were breached, railroads could haul the same loads, as they have always done. Even in Lewiston, where people rabidly backed the dams in the 1960s and onward, support for removal has grown because the uppermost reservoir, lapping at town's edge, constantly aggrades with dam-induced silt deposits. The reservoir is rising between its constraining levees and promises to overtop them, flooding lower elevations of the town if the problem is not addressed.

Touting economic as well as fishery benefits, the Save Our Wild Salmon Coalition argued for dam removal, which in 2011 the American Fisheries Society overwhelmingly endorsed. The same year, Judge James Redden ruled that the federal government's position regarding endangered salmon was flawed. For the fourth time the court ordered preparation of a better plan. However, a new draft in 2014 offered no change.

Ultimately, survival of the Snake River's wild salmon and steelhead depends on politics, thus far controlled by the subsidized interests in dams, hydropower, and industrial barging. The outcome of this controversy will be either a high point in the history of river restoration or one of the greatest tragedies in the loss of iconic wildlife species in America, much like the passing of buffalo on the Great Plains and extinction of the passenger pigeon. Except that now, we know better.

Like the old habit of dumping waste into waterways, taking water out with little regard for the life of the streams was deeply rooted in the culture of Oregon and the West—and still is. Irrigation accounts for 88 percent of Oregon water withdrawals, and 70 percent of that is for pasture and cattle feed. Some diversions are negligible; others leave waterways depleted or bone-dry in summer. Invasive carp, catfish, and bass take over waters that are warmed by the loss of flow, which also causes groundwater to decline and riparian zones to wither.

This status quo is maintained by western water law, long supporting the right of water users to claim any amount they could "beneficially" use. In 1955 the legislature passed a Minimum Perennial Streamflow Act, allowing the state to claim unappropriated water (in the few places where it still existed) for the life of the streams. In 1987 the law was upgraded to allow water-right holders to convert their allotments to in-stream uses rather than simply lose undiverted water to other irrigators. By 2015 Oregon had fifteen hundred designated minimum-flow reservations. This was good compared to other western states, but the deck was still stacked mightily against the streams because the "minimum" flows were calculated to barely maintain life—clearly inadequate to support the full complement of fish let alone the greater riparian community. And virtually all the in-stream rights were secondary to "senior" rights held by irrigators who continued to get all the water they claimed no matter what the annual level of runoff. In 2001 the Oregon Water Resources Department recognized that there "is not enough water where it is needed, when it is needed, to satisfy both existing and future water uses."

While water pollution had been the battleground in river conservation from the 1930s through 1960s, and dam fights defined this movement through the 1970s and 1980s, keeping water in our streams and reclaiming them from lost flows became the core of the river protection movement in the early 2000s. WaterWatch of Oregon led the way in campaigns to prevent depletion of the McKenzie, Deschutes, Kilchis, Clackamas, Row, and other streams.

The hazards that rivers and their native life face in the twenty-first century have been aggravated by a host of unwelcome alien species, introduced accidentally and intentionally from other parts of the globe or nation. Exotics spread without population checks that predators might have maintained in lands of origin.

New Zealand mud snails, for example, multiplied in coastal watersheds and usurped habitat of native bottom life. Along many riverfronts, native willows and cottonwoods were displaced by thickets of Asian knotweed, Scotch broom, and gorse. Some private landowners and public-land stewards had the wherewithal to fight the exotics by pulling or poisoning them, but across much of Oregon, alien species run amok.

Fish hatcheries unfortunately introduce many of the problems of exotic species. Built with the goal of increasing fishing catches, the hatcheries impinge on native and wild fish in fatal ways.

Hatchery-bred fish compete with natives for food and living space. In feedlot conditions, the hatchery stocks are prone to infection, and then they broadcast their diseases to wild fish. The hatchery fish are less capable than the natives at surviving in the streams and in the ocean, but they crossbreed and thereby weaken the wild fish gene pool. Truckloads of hatchery fish dumped into rivers mask the decline of native fish, whose extinction in some cases has loomed without managers even knowing about it.

Solutions are elusive, as the hatcheries maintain a strong constituency among sport and commercial fishermen who don't distinguish between self-sustaining and man-made populations. But Montana's experience is instructive: knowing that their facilities were undercutting wild fish populations, the state closed its hatcheries in 1972. Since then, the fishing has greatly improved. Montana became the most-desired angling state in the nation, and the sport-fishing industry there grew to underpin the entire region's economy.

On select Oregon rivers, the Department of Fish and Wildlife emphasized the survival and management of native fish, but hatcheries—run at great expense—continued to pose threats to wild fish in many—if not most—streams. See biologist Jim Lichatowich's work in *Salmon Without Rivers* and *Salmon, People, and Place* for historical and biological perspectives on hatcheries. The Native Fish Society and Trout Unlimited work for a better balance between hatchery production and wild fish.

The watery paths that rivers take across Oregon's landscapes might look natural to the casual eye, but changes have been profound, with insidious effects on stream life. For example, logjams at first glance might simply look like debris. But in fact, they are vital to river health. They offer structure where insects and other invertebrates cling and nest, and these arthropods subsequently make food for fish. The blockage slows the water enough that gravel is deposited into pebbly piles—perfect spawning beds for fish. The backed-up water infiltrates underground water tables to supply floodplain forests, then seeps back into the river when it's most needed, after surface flows have dropped. The logjams force water to flow around the stranded timber, adding to channel complexity and choices of habitat: deep languid sloughs, shallow fast connectors, intimate backwaters that protect smaller fish, and flooded swamps, where the overflowing water offers shelter for fish eggs and fry and also gathers food in the form of leaves, seeds, and nuts flushed into the aquatic system.

However, most logjams were removed to float logs to sawmills, or to streamline flows and avoid floods on low-lying property, or to open passages for fish—a misguided effort that failed to recognize that fish negotiated most logjams without aid. Lacking logs as barriers, the streams cut unimpeded through soft beds and gravel bars that had been laid down for ages behind the historic blockages, and then the currents entrenched further down to bedrock in rivers such as the Siuslaw, Smith, and Wilson throughout Oregon's timber regions. In effect, the streams became channelized.

Efforts are underway to reinstate some logjams, or at least to allow their natural formation as trees grow tall again and eventually fall into the streams to replicate conditions that existed before the logging era. This requires that protections against logging in riparian zones of federal land remain effective, and that state rules governing riverfront cutting on industrial land are upgraded to allow trees along the rivers to once again grow to adequate size; for making logjams, big trees are needed.

While the loss of healthy channels in timber country was an unintended consequence of logging, channelization in other streams through farms and cities was deliberate, calculated, and systematic. The Willamette once pulsed northward in a riverbed of remarkable variety, fingering with multiple sloughs, meandering as a riparian ribbon eight miles wide in places. But to expand crop acreage, farmers commandeered floodplains by plugging channels, pushing up levees, and armoring the outsides of bends with rock and junked cars. Incrementally narrowed and tamed over the course of a century, the Willamette shrunk to a relatively narrow single channel. Between Eugene and Albany the river's length was shortened by 53 percent.

Within this simplified channel, habitat for fish and wildlife disappeared. Cold water in deep holes of shaded sloughs needed by cutthroat trout became scarce. Virtually unnoticed, the Oregon chub, requiring sluggish backwaters with vegetation, was reduced to endangered status by 1993. In 2014 the pretty little olive-backed minnow became America's first fish recommended for de-listing under the Endangered Species Act. This ostensible recovery, however, owed to reintroductions in tributaries and discovery of formerly unknown populations rather than widespread reinstatement of scarce habitat.

Consequences of channelizing the Willamette went far beyond the loss of chubs, cutthroats, and Chinook. Instead of spreading out, infiltrating groundwater, and pausing in its surge downstream, flood flows in constrained channels speeded up and climbed the walls of narrowed passages to pose ever-greater threats to farms and cities below. Wars of escalating levee-height were fought to determine which side of the river would flood. Once the opposing sides achieved parity in that battle, people upstream and down became the losers because the displaced water rose higher there. Conditions worsened as the constrained riverbed adhered to unexpected axioms of hydrology; it perversely aggraded, or rose in elevation as the current deposited more bed load than it took away. The levees, in effect, become lower relative to the volume of water flowing between them, inevitably becoming less effective and more vulnerable.

After the 1996 Willamette flood, the Western Rivers Conservancy commissioned hydrologist Philip Williams, who determined that restoration of fifty thousand acres of riparian forest could reduce flood levels 18 percent, benefitting the entire downstream corridor, including Portland, without building a single new dam or levee. But the group found few farmers willing to sell land or easements for seasonal overflow. Prospects may change as global warming brings higher and more frequent floods that will rupture levees long considered reliable.

While great progress has been made in river conservation, the largest reason for ongoing losses continues relatively unchecked: floodplain development. On undeveloped floodplains, buffers of open space filter out pollution, for free. Forests shade rivers and keep them cool. Wildlife depend on riparian habitat. High water resupplies groundwater deposits. Floodplains are important as the waterfront where people walk, bike, and fish. But because of development in these lowlands—and in spite of billions of dollars spent on flood control—property losses from river flooding grew to average $8.2 billion annually nationwide in 2013, and that figure is increasing, according to the National Weather Service.

To address the root of this problem by avoiding floodplain development and protecting open space, important riverfront tracts have been bought by government agencies and land trusts. Though the program stalled before completion, parklands of the Willamette Greenway were bought by the state in the 1970s, and dozens of other parcels were acquired by agencies and conservancies along Oregon waterways from Hells Canyon of the Snake River to the mouth of the Sixes at Cape Blanco, bought for a state park in 1971. But the amount of land that can be acquired will always be limited. It's expensive, and many property owners have no interest in selling.

In contrast, floodplain zoning applies to all riverfront landowners in the municipalities that adopt regulations. Though controversial, the authority of local or county governments to zone has been well established by the legislature and courts from the 1920s onward: development of floodplains affects not just landowners but also adjacent owners, downstream properties, the general public, and all levels of government. The benefits of floodplain development are enjoyed by a few, but the costs are borne by all in this stark example of privatizing benefits and communizing costs.

Some form of floodplain zoning exists throughout counties statewide because the Federal Flood Insurance Program mandates it for federally insured home mortgages. Those requirements were intended to ban development in high hazard areas and deep water, but they allow building on fill that elevates structures above the hundred-year frequency flood, and they permit "flood proofing"—dubious specifications that offer no assurance against damages, not to mention protection for rivers or open space. Zoning is prescribed for the hundred-year frequency floods, but floods occur with uncanny frequency, and flooding along many rivers has worsened with watershed disturbance, constriction of nearby levees, and development on fill that—like a rock dropped into a bucket already full of water—pushes water higher elsewhere. Researchers Jon Kusler and Larry Larson found that 31 percent of damage claims under the Flood Insurance Program occurred above the hundred-year floodplain. Oregon's pioneering program in statewide land-use regulations in the 1970s encouraged local governments to exceed the compromised requirements of the Insurance Program, but few do it, and the state itself has failed to pursue this wise policy.

As the twenty-first century unfolded, better zoning appeared to offer the greatest potential for river conservation at the least cost, yet most conservation groups avoided this approach because it involved the nettlesome issues of land-use regulation by local governments. Meanwhile, the specter of climate change looms large, and the inevitable increase in size and frequency of floods is already evident.

In 2013, *Climate Change in the Northwest* reported that a rise of 2 to 8.5 degrees Fahrenheit in air temperatures is expected by the year 2070. Even the minimum would be catastrophic, and just a year later the Climate Change Research Institute of Oregon State University predicted that a rise of seven to fourteen degrees is most likely. The *Rogue Basin Climate Change Impact Report* by the Institute for Sustainable Environment at the University of Oregon projected a searing increase of fifteen degrees by 2080. One estimate called for summers to be 34 percent drier than 2012, with a sharp drop in summer stream levels.

In 2014 the Climate Impacts Research Consortium at OSU reported that the "snow dominated" area of the Northwest might contract 30 percent by 2050, and with it, summertime runoff. An immediate preview of that grim future was evident just a year later, when Oregon had relatively normal precipitation but record-low snowpacks—some of them only *half* of the previous low.

Global warming means that less water will be available but irrigation demands will increase. With higher temperatures and intensified storms, winter floods will be higher. In summertime, waters will heat, threatening already stressed cold-water trout, steelhead, and salmon. *Proceedings of the National Academy of Sciences* reported in August 2011 that western trout habitat could decline 50 percent by 2080 because of global warming.

Sea level will rise four to fifty-six inches by 2100, which will cause catastrophic shoreline erosion and saltwater encroachment up estuaries and rivers. Forest fires will intensify, as recent years attest. In their wake, silt will be flushed into rivers. Without shade on the charred landscape, water temperatures will climb further.

For decades, river conservationists have sought to establish open space buffers along waterways to protect riparian habitat, to benefit wildlife by linking existing public open spaces along rivers, and to maintain cool water temperatures and flows by limiting the amount of water that's diverted. Because of the climate crisis, those tasks will have to be done more frequently, better, and faster.

Looking for a positive alternative to problems that afflict rivers, conservationists persuaded Congress to pass the National Wild and Scenic Rivers Act in 1968, and the rivers of Oregon were crucial to the establishment and growth of that program. Through legislation, specified reaches can be designated for Wild and Scenic status, which bans dams, hydropower approvals, and other federal actions that would harm the river. Modeled on the national system, in 1970 Oregon voters, by a two-to-one margin, established a State Scenic Waterways program of seventeen streams, later expanded to twenty-four, with more additions possible.

The Rogue's eighty-five-mile lower canyon was among the prestigious rivers in the original national program. In 1975 Congress added the Snake for sixty-seven miles in Hells Canyon. In 1984, the lower Illinois was added, including a dam site once proposed at Buzzard's Roost.

Then a 1988 bill transfigured the National Wild and Scenic system. Starting with Forest Service studies that identified potential Wild and Scenic mileage, the Oregon Rivers Council (now Pacific Rivers) and others succeeded in getting forty-four rivers plus nine tributaries designated for 1,442 miles—the largest number of rivers ever added to the National Wild and Scenic system at one time.

Six years later, eleven miles of the Klamath were included to block the City of Klamath Falls from diverting the river for hydropower. In 2009, portions of another nine small streams in the Mount Hood area were included, plus two forks of the Elk at the South Coast. In 2015, Oregon had seventy-one rivers and named tributaries in the Wild and Scenic system—second highest in the nation—and 1,931 designated miles—third highest. Though significant, this was only 1.7 percent of the state's total stream miles. The national rivers were a cherished group, but myriad tasks remained to protect and restore the rivers of Oregon.

Thousands of Oregonians in dozens of organizations are addressing the ongoing threats to our rivers. In a unique approach to prevent salmon from going extinct (and to avoid official federal listing of some runs as threatened or endangered), the Oregon Plan for Salmon and Watersheds aims "to restore native fish populations and the aquatic systems that support them." The program invested $240 million in restoration between 1999 and 2015. Through ninety-one watershed councils, trees are planted along riverbanks to cool the water, logs are placed in streams to reinstate salmon habitat, fences are built to steer cattle away from eroding banks, and culverts are upgraded to allow fish passage.

WaterWatch leads the way toward restoring flows and reclaiming streams degraded by diversions. The Native Fish Society works for wild salmon, steelhead, and cutthroat. Trout Unlimited labors for clean water and habitat. American Rivers strives to protect the best rivers from mining and other threats, and to add them to the National Wild and Scenic system. Willamette Riverkeeper aims to halt pollution and clean up toxic refuse, as do Riverkeeper organizations of the Columbia, Rogue, Coos, Klamath, and Tualatin. Pacific Rivers Council seeks to sustain stream protections of the Northwest Forest Plan and to improve logging practices on industrial forest land. American Whitewater does whatever is needed to protect the finest paddling rivers. The Western Rivers Conservancy buys riverfronts for open space, as do The Nature Conservancy and local land trusts. Oregon Wild fights for rivers flowing through federal lands. Klamath-Siskiyou Wildlands Center wants to safeguard the incomparable suite of streams from the Smith at Oregon's southern border north through the Coquille. The Hells Canyon Preservation Council watches over Oregon's deepest canyon. Similar groups have adopted other regions.

River protection has been a high priority among Oregonians since the 1938 initiative to clean up waterways and the 1970 vote to establish the Oregon State Scenic Waterways Program. In 2013 the Oregon Values and Beliefs Survey by Oregon Health and Science University found that—even in hard economic times with a din of anti-regulation rhetoric—57 percent of Oregonians believed that protecting the environment should be given priority even at risk of slowing economic development.

The history of river conservation in Oregon has come a long way from the first campaigns for clean water. The beauty and health of the natural streams that remain can inspire people to create a future in which rivers continue to flow as the essence of Oregon.

Chetco River and coltsfoot

Quiet woodland pools along the Chetco nourish a luxuriant crop of coltsfoot and red alders in the upper river gorge.

CHAPTER 5   The Rivers Around Us

At the beginning of this book, I reflected on my journey down the Willamette River, gently winding through the heart of Oregon's most populated valley. Plentiful flows whisked me past shores constrained by farmland but still green with cottonwoods. The riparian artistry that I found will enchant anyone who drifts with the current and notices what's there.

Here at the end of the book, for a counterpoint to the Willamette's welcoming ease of travel, I journeyed with my wife, Ann, and a small group of friends to one of our least tamed and most remote rivers, the Chetco.

While upper reaches of that coastal stream are all wilderness, and visiting requires a long hike, an isolated mid-section can be reached by a rough road leading to a remote riverfront. There we unpacked our inflatable kayaks, strapped a waterproof bag of supplies onto each, and began to paddle—not downriver, but up. We wanted to see the nature of an extraordinary canyon, and our boats offered the only way to go.

We paddled through pools that were crystalline and so deep we could not dive to the bottom. Then we dragged our gear over boulders and gravel bars up to the next pool, again and again. We passed the mouths of tributaries feeding pristine waters into the river. We encountered cliffs and tangles of fallen cedars as we pressed onward for several miles to Boulder Creek and the Kalmiopsis Wilderness—among the first Wilderness Areas designated in America and the third largest in Oregon. We camped on the gentle slope of a gravel bar where evergreens shaded our tents and where the Chetco shimmered past, its music entertaining us from rapids both above and below. While the Willamette had offered me a comforting tour through Oregon's heartland with farms and towns along the way, the Chetco was as wild as it gets.

The next day we began our descent, passing our put-in point and continuing downstream for another two days through lower canyons. We dropped over dozens of rapids that would have been threatening in the flush of springtime, but on low summertime runoff we slipped through, or bounced off rocks in our forgiving air-filled craft, or clambered over slippery rocks, lugging our gear and pairing up to push and pull our boats.

Two impressive bottlenecks required portaging over boulders the size of cars and buses. The first was an extravaganza of maroon, white, and gray-striped rocks in a rapid called Radiolaria for the microscopic creatures that had fossilized to create those rocks undersea prior to the Coast Range uplift. Then Conehead's radical drop—shaded by two pointed monoliths and clogged with logs—demanded that we line our craft through narrow chutes and a slot barely boat-wide.

Downstream from that last canyon, a more casual Chetco is easily reached along a forest road. Its refreshing riffles and deep green pools—alive with spawning salmon in winter and swimming children in summer—are no less beautiful than the wilderness above. It's a riverfront, like many in Oregon, where you can see a free-flowing waterway connecting mountains to sea, forests to fish, and people to the natural world around them.

For other wild scenes and remarkable river adventures in Oregon I could have gone to the Salmonberry, North

Chetco River "portage"
With a flow of only fifty cubic feet per second, we were able to cautiously wade and pass our boats down Conehead Rapid, which at higher flows would have posed a treacherous tangle of undercut rocks and logs.

Umpqua, or Collawash; the Metolius, Imnaha, or Wenaha. Whether we explore these or the Willamette—or any of a hundred rivers and thousand streams in between—Oregon is a wonderland of waterways. The rivers are all around us. See for yourself. Start today in your own backyard. Where does that water come from? Where does it go?

   Adopt a river as your own. Share the knowledge you gain with your family and friends. Through many decades of history, Oregonians have proven that we have the power to protect what remains, to restore what has been lost, and to keep intact these natural gifts that will be needed by the generations to come.

Rogue River near Lost Creek
A fresh winter snowfall whitens elegant shores of the upper Rogue.

Ramona Falls
Hidden in the wilds of Mount Hood's western flank, a tributary to the upper Sandy sprays onto vertical columns of basalt.

McKenzie River above Deer Creek
Vine maples green the shores of the swift upper McKenzie.

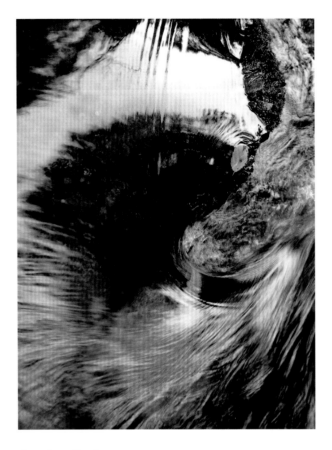

Panther Creek

This tributary to the Elk River swirls over bedrock at the outlet of a pool where salmon and steelhead spawn.

Wychus Creek

Chush Falls sprays onto basalt outcrops and moss-covered logs in the Three Sisters Wilderness.

Hubbard Creek

Hubbard Creek completes the hydrologic cycle as it flows to its end in the Pacific Ocean.

ACKNOWLEDGMENTS

Editor Mary Elizabeth Braun of Oregon State University Press saw the value of my book and championed its passage through publishing, all the time glowing with friendship. Sincere thanks also to book designer Erin Kirk New, and to Tom Booth, Micki Reaman, and Marty Brown at the Press. Oregon is a better place because of their work. Steve Gordon of Cartagram did a masterful job of producing the map.

For financial support to aid the Press in this ambitious leap and to cover a few of my costs, I'm deeply grateful to Bert Kerstetter. Also to Adam Smith, Nelson Smith, Melanie Smith, and all the trustees of the Whole Systems Foundation; to Becky and Steve Schmitz; and to Greg Knight and Marieta Staneva of the Knight-Staneva Foundation for Sustainability and Future Environments.

I offer special recognition here to Greg Knight—a distinguished university professor who dedicated a lifetime to geographic work, to his students, and to our understanding of water resources. He engaged with passion in the important issues of his day and, to the very end, remained a heartwarming friend. He gave generously of his time, intellect, and life, and in those ways inspired others to do the same.

Travis Hussey was the best paddling buddy I can imagine. Nat Hart reviewed my manuscript with a well-trained mind for language. Dave Fortney and Steve Thompson offered tips about digital photography. I delighted in sharing river-exploring enthusiasm with Zach Collier from Northwest Rafting Company. For early inspiration, thanks to my old friend and the dean of Oregon river running, Bob Pierce, who invited me to join his group for my very first Oregon river journey in 1977.

Stan Gregory and Doug Markle of Oregon State University and biologist Jeff Dose formerly of the Forest Service provided sharp insight about fish.

Our rivers would be greatly diminished without the conservation work of John DeVoe of WaterWatch, Jack Williams of Trout Unlimited, Bill Baake of Native Fish Society, Phil Wallin and Sue Doroff of Western Rivers Conservancy, Travis Williams of Klamath Riverkeeper, Tom O'Keefe of American Whitewater, David Moryc of American Rivers, John Kober of Pacific Rivers Council, Don Elder formerly of River Network, Forrest English of Rogue Riverkeeper, Laurie Pavey of the Willamette Kayak and Canoe Club, and a host of others working and volunteering for these and other groups. All were helpful to me.

For information of many kinds, I depended on the capable authors listed under Sources. For following a similar path two decades before me, special recognition here goes to the superb photographer Larry Olson and his now out-of-print book of great elegance about rivers in the National Wild and Scenic Rivers system: *Oregon Rivers*.

The greatest pleasures were river trips with my wife, Ann Vileisis, who—while writing a challenging book of her own—provided enthusiasm, brilliance, editing, time, patience, support, love, and tireless work in her many gifted ways to protect the rivers where we live along the coast of southern Oregon.

SOURCES

This is a short list of books and articles that I found helpful. For more, see the sources in *Field Guide to Oregon Rivers*.

Bastasch, Rick. 1998. *The Oregon Water Handbook*. Oregon State University Press.

Defenders of Wildlife, et al. 1998. *Oregon's Living Landscape*.

Frissell, Christopher A. 2013. *Aquatic Resource Protections in the Northwest Forest Plan: Evaluating Potential Consequences of Proposed Riparian Reserve Reductions for Clean Water, Streams and Fish*. Coast Range Association. coastrange.org.

Hawley, Steven. 2011. *Recovering A Lost River: Removing Dams, Rewilding Salmon, Revitalizing Communities*. Beacon Press.

Huntington, Charles, Willa Nehlsen, Jon Bowers. March 1996. "A Survey of Healthy Native Stocks of Anadromous Salmonids in the Pacific Northwest and California." *Fisheries*.

Keller, Robb. 1998. *Paddling Oregon*. Falcon.

Kerr, Andy. 2004. *Oregon Wild*. Timber Press.

Lichatowich, Jim. 2013. *Salmon, People, and Place*. Oregon State University Press.

_____. 1999. *Salmon Without Rivers*. Island Press.

Olson, Larry N. and John Daniel. 1997. *Oregon Rivers*. Westcliffe.

Oregon Department of Fish and Wildlife. 2009. *Fishing, Hunting, Wildlife Viewing, and Shellfishing in Oregon*.

_____. 2012. *Oregon Conservation Strategy*.

_____. 1993. *Oregon Wildlife Diversity Plan*.

_____. 2010. *Oregon Native Fish Status Report*.

Oregon Department of Water Resources. 2010. *Integrated Water Resources Strategy*.

Orr, Elizabeth L. and William Orr. 2012. *Oregon Geology*. Oregon State University Press.

_____. 2005. *Oregon Water: An Environmental History*. Inkwater Press.

Palmer, Tim. 1997. *The Columbia*. The Mountaineers.

_____. 2014. *Field Guide to Oregon Rivers*. Oregon State University Press.

_____. 1991. *The Snake River: Window to the West*. Island Press.

_____. 1993. *The Wild and Scenic Rivers of America*. Island Press.

Reeves, Gordon H., Kelly M. Burnett, Stanley V. Gregory. 2002. "Fish and Aquatic Ecosystems of the Oregon Coast Range," in *Forest and Stream Management in the Oregon Coast Range*, Oregon State University Press.

US Forest Service, Pacific Northwest Research Station. 1997. *Highlighted Scientific Findings of the Interior Columbia Basin Ecosystem Management Project*.

Palmer, Tim and Ann Vileisis. 2008. *The Great Rivers of the West*. Western Rivers Conservancy. westernrivers.org/about/greatrivers/

Wildlife Society. 1994. *Interim Protection for Late-Successional Forests, Fisheries, and Watersheds*.

Willamette Kayak and Canoe Club. 2004 and previous editions. *Soggy Sneakers*. Mountaineers Books.

Yuskavitch, Jim. 2001. *Fishing Oregon*. Falcon.

ABOUT THE AUTHOR AND PHOTOGRAPHER

Tim Palmer
photo by Ann Vileisis

Tim Palmer has written twenty-four books about rivers, conservation, and adventure travel. He has spent a lifetime canoeing, rafting, hiking, exploring, and photographing along streams, and his passion has been to write and speak on behalf of river conservation. He has paddled on more than 350 rivers nationwide, including 50 in Oregon.

Recognizing his contributions in writing, photography, and activism, American Rivers gave Tim its first Lifetime Achievement Award in 1988, and Perception Inc. honored him as America's River Conservationist of the Year in 2000. California's Friends of the River recognized him with both its highest honors, the Peter Behr Award and the Mark Dubois Award. *Paddler* magazine named Tim one of the "ten greatest river conservationists of our time," and in 2000 included him as one of the "hundred greatest paddlers of the century." In 2005 Tim received the Distinguished Alumni Award from the College of Arts and Architecture at Pennsylvania State University. Topping off these honors, he received the National Conservation Achievement Award ("Connie") for communications given by the National Wildlife Federation in 2011.

Tim's *Rivers of America* was published by Harry N. Abrams and features two hundred color photos of streams nationwide. He also photographed and wrote *Rivers of California*. *The Columbia* won the National Outdoor Book Award in 1998, and *California Wild* received the Benjamin Franklin Award as the best book on nature and the environment in 2004. *The Heart of America: Our Landscape, Our Future* won the Independent Publisher's Book Award as the best essay and travel book in 2000. In 2015 his *Field Guide to Oregon Rivers* was a finalist for the Oregon Book Award and for *Foreword Review*'s Adventure Book of the Year and was the winner of the National Outdoor Book Award.

Tim lives in Port Orford and has explored rivers on foot and by canoe and raft throughout the state.

Before becoming a full-time writer and river conservation activist, Tim worked for eight years as a land-use planner. He has a bachelor of science degree in landscape architecture and is a Visiting Scholar with the Department of Geography at Portland State University and an Associate of the Riparia Center in the Department of Geography at Pennsylvania State University. He speaks and gives slide shows for universities, conservation groups, and conferences nationwide. See his work at www.timpalmer.org.

ABOUT THE PHOTOGRAPHS

For many years I used a Canon A-1 camera with 17-200 mm FD lenses, but most of the photos here were taken with a Canon 5D digital camera with 17-200 L series zoom lenses and a 50 mm L series lens. I also use a Canon underwater Powershot for an occasional view of what's happening at the waterline or beneath it. For backpacking and other adventures when a small kit is needed, I carry a Fujifilm digital X-E2 with its superb 18-55 and 55-200 XF zoom lenses.

With the goal of showing nature accurately and realistically, I limit myself to minor adjustments for contrast and color under Apple's very basic iPhoto program. I use no artificial light or filters, and do nothing to alter the content of the photos.

I don't dress my photos up, but I do search for the most beautiful scenes I can find. Most pictures were taken in early morning or evening with nature's elegant low light that's full of color, shading, and shadows. Using a tripod, I take long exposures to keep my ISO low and to maximize depth of field, which also gives flowing water the streamed effect of motion. I vary perspectives from high to low, climb trees, shoot from the canoe and raft, wade shallow or deep, and slip underwater now and then.

My ultimate operating principle is "be there, be there, be there." No amount of gear, technique, or artistry can take the place of knowing the subject well and spending time intimately and receptively wherever the waters flow.

INDEX

alder trees, 16
Alsea River, iv, 7, 31, 136
American Fisheries Society, 142
American Rivers (organization), 147
Applegate Dam, 141
Baker Creek, 140
Bear Creek (Rogue basin), 62
bull trout, 74, 86, 107, 124, 127, 133, 138
Cape Creek, 33
channelization, 36, 144
Chetco River, 6, 42, 43, 148–150
Chewaucan River, 117
Clackamas River, 86, 87
Clearwater River, 48
climate change, 146, 157, 158
climate, 9
coastal rivers, 21–43
Collawash River, 88
Columbia Gorge, 90–94
Columbia River, 10, 90–93, 119, 121, 141, 142
dams, 142
Coquille River, South Fork, 37, 139
cottonwood trees, 14, 15, 17
Cow Creek, 53–55
Crooked River, 108
Cummins Creek, 33
cutthroat trout, 12, 13, 33, 97, 137, 142, 144
dams, 3, 6, 10, 11, 13, 14, 17, 20, 68, 74, 87, 93, 103, 108, 115, 118, 137, 140–142

fights over, 6, 74, 119, 120, 140–143, 147
removal of, 56, 62, 116, 121, 142
dam-free rivers, 2, 20, 33, 43, 44, 82, 97, 100, 110
Deschutes River, 100–103
diversions, 103, 108, 137, 142, 143
Douglas, Justice William O., 141
Drift Creek (Alsea basin), 7, 136
Eagle Creek (Columbia Gorge), 94
East Lostine River, 132, 134
Elk Creek Dam, 141, 142
Elk River, 14, 18, 39, 41, 154
estuaries, 33, 36, 38, 42, 44
Evans, Brock, 141
fish, 2, 9–14, 44, 137, 138, 140, 144. *See specific rivers.*
floods, floodplains, 13, 14, 17, 145, 146
Flora Dell Creek, 20
Forest Practices Act, 139, 140
Furnish, Jim, 24
geology, 9, 10
global warming, 146
Grande Ronde River, 128–130
hatcheries, 110, 143, 144
Hells Canyon Preservation Council, 147
herbicides, 140
Hood River, West Fork, 94
Horsetail Falls, 92
Hubbard Creek, 42, 156
Hussey, Travis, 7, 21, 157

hydroelectric power, 74, 93, 140–142, 146, 147
Illinois River, 62
Imnaha River, 126, 127
irrigation, 56, 140, 142, 146
John Day River, 10, 110–114
John Day River, Middle Fork, 114
John Day River, North Fork, 110, 111
Kilchis River, 26
Klamath River, 115, 116, 138, 147
Klamath Siskiyou Wildlands Center, 147
landslides, 139, 140
Leopold, Aldo, 17
levees, 145
Lichatowich, Jim, 144
Little North Santiam River, 78
logging, 24, 44, 138, 139, 140, 144
logjams, 22, 41, 90, 99, 103, 144
Lostine River, East, 132, 134
Malheur River, 124
McCall, Tom, 138
McKenzie River, iv, 2, 72–77, 141, 143, 152
Metolius River, 5, 106, 107
Minam River, 132
mining, 62, 147
Mollala River, Table Rock Fork, 82, 83
National Creek, 59
Native Fish Society, 144, 147
Nature Conservancy, 147
Nehalem River, 23–25, 33
Nestucca River, iv, 29

161

North Santiam River, 16, 79
North Umpqua River, 46–48
Northwest Forest Plan, 140
Oneonta Creek, 90, 91
Opal Creek Wilderness, 80
Oregon chub, 144
Oregon Plan for Salmon and Watersheds, 147
Oregon State Scenic Waterways Program, 147
Oregon Values and Beliefs Survey, 147
Oregon Wild, 142, 147
Owyhee River, 123
Pacific Rivers (organization), 147
Panther Creek, 41, 154
Paulina Creek, 104, 105
Pine Creek (Snake basin), 125
pollution, 115, 138, 140, 143, 145, 147
restoration, 36, 56, 62, 82, 116, 117, 121, 138, 142, 145, 147
riparian habitat, 13, 145
river systems of Oregon, 10
Roaring River, 89
Rock Creek (coastal), 33
Rogue River, 4, 8, 20, 56–58, 62, 70–74, 141, 142, 146, 151
Rogue River, Middle Fork, 60
Rogue River, South Fork, 61
Rogue Riverkeeper, 147

Salmon River (coast), 29
Salmon River (Idaho), 119
Salmon River (Sandy basin), 99
salmon, 11, 12, 24, 41, 93, 110, 121, 138, 144, 147. *See specific rivers.*
Salmonberry River, 25
Sand Canyon Creek, 99
Sandy River, 96, 97, 99, 142, 152
Santiam River, 78
Santiam River, North, 16, 79
Santiam River, South, 78
Save Our Wild Salmon Coalition, 121
Siletz River, 30
Silver Creek, North Fork, 84, 85
Siuslaw River, xii, 21–24, 33
Sixes River, 38
Smith River (Umpqua basin), 36
Snake River, 6, 118–121, 141–142, 146
South Santiam River, 78
South Umpqua River, 15, 33, 50, 52, 141
South Umpqua River, iv, 15, 33, 44, 45, 50, 52, 141
steelhead, 11-13, 24, 46, 82, 140
sturgeon, 13, 140
Susan Creek, 48
Sweet Creek, iv, 35
Tenmile Creek (Reedsport), 36
trails, 33, 35, 36, 46, 53, 71, 73, 76, 85, 86, 94, 99, 103, 124, 132, 134

Trout Unlimited, 36, 144, 147
Umpqua River, 12, 13, 44–45
Umpqua River, North, 46–48
Umpqua River, South, iv, 15, 33, 44, 45, 50, 52, 141
Wallowa Mountains rivers, 134
Wallowa River, West Fork, 131
waterfalls, v, 3, 23, 35, 47, 48, 56, 59, 68, 73, 85, 92, 100, 105, 106, 109, 141, 152, 154
WaterWatch of Oregon, 143, 147
Wenaha River, 133
Western Rivers Conservancy, 111, 145, 147
White River, 100, 109
Whychus Creek, 106, 154
Wild and Scenic Rivers, 119, 146, 147
wildlife, 13, 14, 17, 64, 82, 117, 142–146
Willamette River, North Fork of Middle Fork, 71
Willamette River, 1-3, 10, 64–69, 110, 137, 138, 141, 144, 145, 149
Willamette River, Coast Fork, 35
Willamette Riverkeeper, 147
Willamette, Middle Fork, 71, 74
Williamson River, iv
Wilson River, 27
Yamhill River, 81
Zigzag River, 99